THE CFZ YEARBOOK
2003

Edited by Jonathan Downes and Richard Freeman

Typeset by Jonathan Downes,
Cover and Layout by Hennis/I am Kurious Oranj for CFZ Communications
Using Microsoft Word 2000, Microsoft , Publisher 2000, Adobe Photoshop CS.

Photographs © 2008 CFZ except where noted

First published in Great Britain by CFZ Press

**CFZ Press
Myrtle Cottage
Woolsery
Bideford
North Devon
EX39 5QR**

© CFZ MMVIII

All rights reserved. Without limiting the rights under copyright reserved above, no part of this publication may be reproduced, stored in or introduced into a retrieval system, or transmitted, in any form of by any means (electronic, mechanical, photocopying, recording or otherwise), without the prior written permission of both the copyright owners and the publishers of this book.

ISBN: 978-1-905723-26-3

Contents

- 5. Prologue to the 2008 edition by Jonathan Downes
- 7. Introduction by Jonathan Downes
- 9. *Tracking down Dorset's black dogs* by Mark North
- 27. *If you go down to the woods today: An overview unsung Native American cryptids* by Mike Hallowell
- 35. *Fortean Zoological aspects of the `Rupert the Bear` stories* by Richard Muirhead
- 39. *The monsters of Rupert Bear* by Richard Freeman
- 43. *Cryptozoological reports from the Second World War* by Jonathan Downes and Nick Redfern
- 75. *Terrors of the Taiga: The monsters of Siberia* by Richard Freeman
- 99. *Source material on the 2002 wave of animal mutilations in Argentina, together with an overview of the episodes in question* by Jonathan Downes
- 133. *Kent's exotic cats caught on film* by Neil Arnold
- 145. *`The case of the creeping fox terrier clone` or `how a sea serpent sighting crept back in time`* by Chris Moiser
  Darwinism: A crumbling theory (An overlooked explanation for why the fossil record shows primitive and complex life appearing suddenly on Earth, with no predecessors, is extraterrestrial intervention) by Lloyd Pye
- 175. Director's report for 2002

Contents

- 5. Prologue to the 2008 edition by Jonathan Downes
- 7. Introduction by Jonathan Downes
- 9. *Tracking down Dorset's black dogs* by Mark North
- 27. *If you go down to the woods today: An overview unsung Native American cryptids* by Mike Hallowell
- 35. *Fortean Zoological aspects of the `Rupert the Bear` stories* by Richard Muirhead
- 39. *The monsters of Rupert Bear* by Richard Freeman
- 43. *Cryptozoological reports from the Second World War* by Jonathan Downes and Nick Redfern
- 75. *Terrors of the Taiga: The monsters of Siberia* by Richard Freeman
- 99. *Source material on the 2002 wave of animal mutilations in Argentina, together with an overview of the episodes in question* by Jonathan Downes
- 133. *Kent's exotic cats caught on film* by Neil Arnold
- 145. *`The case of the creeping fox terrier clone` or `how a sea serpent sighting crept back in time`* by Chris Moiser
  Darwinism: A crumbling theory (An overlooked explanation for why the fossil record shows primitive and complex life appearing suddenly on Earth, with no predecessors, is extraterrestrial intervention) by Lloyd Pye
- 175. Director's report for 2002

Prologue to the 2008 reissue

I'm really enjoying overseeing the series of reissues of the CFZ Yearbooks. As one gets older, all the cliches which one has heard all one's life about ageing, sadly, begin to come true, and one begins to forget things. As the time of writing, the CFZ is 17 years old, and an awful lot of water has passed under an awful lot of bridges in the intervening years, and it is embarrassing to realise exactly how much you have forgotten. For example, when we first opened the original electronic files which contained the masters of this volume, I was amazed to see quite what a wide-ranging book the 2003 Yearbook actually was, and - so far in this reissue program -I think that it is my favourite.

I feel particularly sorry for Neil Arnold; the evidence presented in his article is outstanding, and the pictures are some of the most conclusive ever taken of British big cats. However, when one looks back at the canon of literature on the subject, his discoveries and nowhere to be seen. To the best of my knowledge, the 2003 CFZ Yearbook was the only place where his pictures were ever published.

2002/3 was a pivotal time for the CFZ, and it was -in many ways - the year when the organisation that we know today really started to emerge. I think that this Yearbook, with its new-found professionalism, and fascinating range of articles mirrored that, and set the scene for what was to happen in the next few years. Because 2003 was the first year that we carried out a major foreign expedition off our own back, and it was this following on from the series of high-profile UK based expeditions of 2002/3 that really put us on the cryptozoologist map once and for all. After years of me claiming that we were the biggest cryptozoologist organisation the world, suddenly, I wasn't lying any more...

Jon Downes.
Woolsery,
North Devon
Easter Sunday 2008

Introduction

Dear Friends,

Welcome to another Yearbook. When we came up with the idea of putting out an annual compendium of research papers, and academic documents, too lengthy for inclusion in *Animals & Men*, I don't know whether anyone involved truly thought that nearly eight years later we would still be publishing them. However, the CFZ itself has changed massively since those early days. From being a small, but intense, group of hobbyists we are now a large international organisation, and I believe that we can now truly say what we have been saying (with fingers crossed behind our backs), for years now. We ARE the biggest and best Cryptozoological Research Organisation in the world!

We have big plans afoot for the next few years, and as we begin our second decade of existence I would like to thank everyone who has stuck with us for the last ten years, and ask for your continued support in the future. Trust me, it will be worth it.

Slainté Mhôr

Jon Downes
(Director)

Tracking Down Dorset's Black Dogs

by Mark North

Local legends, folklore and phenomena have long fascinated the CFZ artist and cartoonist Mark North. Both these interests, and illustrating, have resulted in his first publication *Dark Dorset Tales of Mystery, Wonder and Terror* - a compendium of over four hundred tales from his home county. More information can be found by visiting his website at www.darkdorset.co.uk.

The author stands outside the Black Dog Inn sign, Uplyme.

Whilst investigating ghosts, especially animal ghosts for the book *Dark Dorset*, I discovered that there appears to be two main types of spectral dog roaming the Dorset countryside: the pack of hounds associated with the wild hunt, and the large solitary black dog.

The black dog being perhaps the most familiar in Dorset, it appears in three distinct incarnations such as an elemental spirit, a guardian and a harbinger of death.

Accounts of spectral black dogs turn up with extraordinary regularity in the British Isles. Devon, Dorset, Somerset, East Anglia, Lincolnshire, Yorkshire and Worcester all boast of having at least one or several Black Dog sightings, some of which date back many centuries. The most famous is the Black Dog of Bungay, which appeared to a terrified congregation of St Mary's Church in 1577, killing two of them as they knelt at prayer, and injuring and disfiguring another.

Though these supernatural creatures differ from county to county with their names, they all share the same qualities. They are usually distinguished from other domesticated dogs by their large appearance - usually the size of a calf - with a black shaggy coat. They have huge, fiery, saucer-shaped eyes and have the strange behaviour of vanishing at will. Black dogs are known to patrol routes that follow ancient pathways and boundaries, or barring entrances to a gate, stile or road to prevent travellers from proceeding further on their journey.

The black dog can take many forms and is often regarded as the Devil, a witch, a fairy, a messenger, a guardian of buried treasure, a protector of lost travellers, a churchyard grim or even Death itself.

The last guise is the most commonly associated with the black dog, and its pedigree can be traced as far back as ancient Egypt, and the much feared jackal-headed God of embalming, Anubis (pronounced Anu-bis) whose task was to take the souls of the dead before the judge of infernal regions. In ancient Greek mythology, the death dog appears too, as the ferocious three-headed dog known as Cerberus (pronounced Sir-ber-us), whose task was to guard the gates of the underworld of Hades, to prevent the dead from escaping. Even in Norse mythology, we have a similar hound, Garmr (pronounced Garm). This was a huge black hound with a blood spattered breast, and eyes of burning coals which watched over the gates of Hell, where he ushered the souls of the dead into the underworld.

Looking up the words black dog in any modern English dictionary, one can find that the definition now means to become depressed or melancholy, as in the expression 'The Touch of the Black Dogs', or in old country sayings, such as 'The Black Dog is at his heels' meant that a person was about to die. Sir Winston Churchill often referred to his moods of depression as his 'black dogs'.

> **black dog**. 1. melancholy, depression of spirits; ill humour. In some country places, when a child is sulking, it is said 'the black dog is on his back'. (**Oxford English Dictionary**)

In Dorset, stories of black dogs that can foretell illness or death are said to haunt the Black

Down around The Hardy Monument, near Portesham, and the road between Drimpton and Broadwindsor at midnight.

However, there are two exceptional recorded incidents of a black dog appearing as a harbinger in Dorset. An elderly lady living in Bridport in 1915 told the first, and earliest account. She recounted how one Sunday after attending church a woman and her companion were making their way home along a quiet lane, when the woman was suddenly jolted by what she described as 'a girt black dog so big as a donkey,' rushing past her. Turning quickly to her friend she said. "What's that?" only to glimpse the creature as it disappeared, leaving the woman profoundly shaken by the experience, and her friend very puzzled. It was not until the woman reached her home that she found her daughter had died during her absence. She understood the dog to have been a portent.

The second, and most recent, recorded account of a harbinger once appeared to a Beaminster woman in 1957. She described seeing a large black dog with long ears and large staring eyes that walked around her several times before disappearing, within two months she was dead.

AQUATIC BLACK DOGS

The black dog's fame has often been greater in parts of the country where it has a name, for example the Black Shuck of East Anglia and the Gytrash of Lancashire. At the southern most tip of the Isle of Portland, near a place known as the Bill, there is a hole close to the cliff edge that forms a natural twenty-foot cavern known as Cave Hole. During extreme stormy weather, it is advisable to keep well away from it, as this is the lair of the dreaded Roy Dog!

This animal is described as a shaggy black dog, as high as a man, with large fiery eyes - one green, one red - and entwined in his mane of dark fur can be seen the freshly plucked eyes of his victims.

It is said that the creature emerges from the watery depths to seize any traveller passing by Cave Hole, and drags them down into his dark watery domain.

The Row Dog of The Isle of Portland - probably another version of the name Roy, and most likely the same creature - appears during the hours of darkness and prowls the island without any particular purpose, though, if you encounter it, you are confronted by a huge black shaggy dog, the size of a man, with large penetrating saucer-shaped eyes. His particular habit is not to attack you, but to merely obstruct your way snarling and barking aggressively.

Suggestions to possibilities behind the origin of the name of the Roy Dog may lie in Anglo-Saxon myth. According to Germanic belief, The Roggenwolf (Rye Wolf or Corn Dog) are spirits of the fields, and the movement of a swathe of ripe corn is attributed to these invisible spirit animals. During harvest-time, the animals would seek to escape the scythe and take refuge in whatever sheaf was left standing. This sheaf held the spirit of the corn till the following year. Therefore, we have a striking similarity in name between the Roy dog and the Rye wolf.

However, the Roy Dog behaves not unlike an etheric and astral elemental or water fairy that

The Roy Dog

The foghorn blasted loudly
As the sea mist suffocated the land
Warning mariners of the monster
The Isle of Portland.

I wasn't far from Cave Hole
Where a fisherman was once found dead
No marks were ever found on his body
But his eyes had been plucked from his head.

Rumours blamed his death on a spectre
A beast called the Roy Dog
That stalks in stormy weather
And roams in silent fog.

Again the Bill's voice blasted
Cutting through the ghostly grey
In the deathly silence that followed
I heard something more sinister than sea spray.

A distant soft pad, pad, padding
An eerie repetitive sound
Closer and closer towards me it came
Which made my heart pound.

I felt the colour drain from my face
As fear began to take hold
Could the legend really be true?
Just the thought of it left me cold.

As the distant padding grew ever nearer
I desperately sought to see through the fog
And emerging from the veil
Came the dreaded Roy Dog!

It was larger than a mortal dog
With sharp fangs eager to rip and tear
Flaming eyes, one red one green
And a black coat beyond compare

But what I think was worse of all
Around its neck was a trophy of fear
All seeing dead men's eyes
That never cry a tear.

The smell of decay crept up my nose
And all of a sudden I felt very cold
Something said my days were done
At this sight that I behold.

I made the sign of the cross
And closed my eyes ready to meet my fate
And like a statue there I stood
And for a long time I did wait.

At last curiosity got the better of me
And slowly I opened one eye
I found myself all alone
For the creature I could not spy.

Now why did the Roy Dog spare me?
I would really wish to know
Was it the sign of the cross I made?
Which made the creature go.

But I lived to tell you the tale
To tell you to keep out of the fog
When the Bill sounds its warning
Beware the Roy Dog!

By Robert Newland, 2002

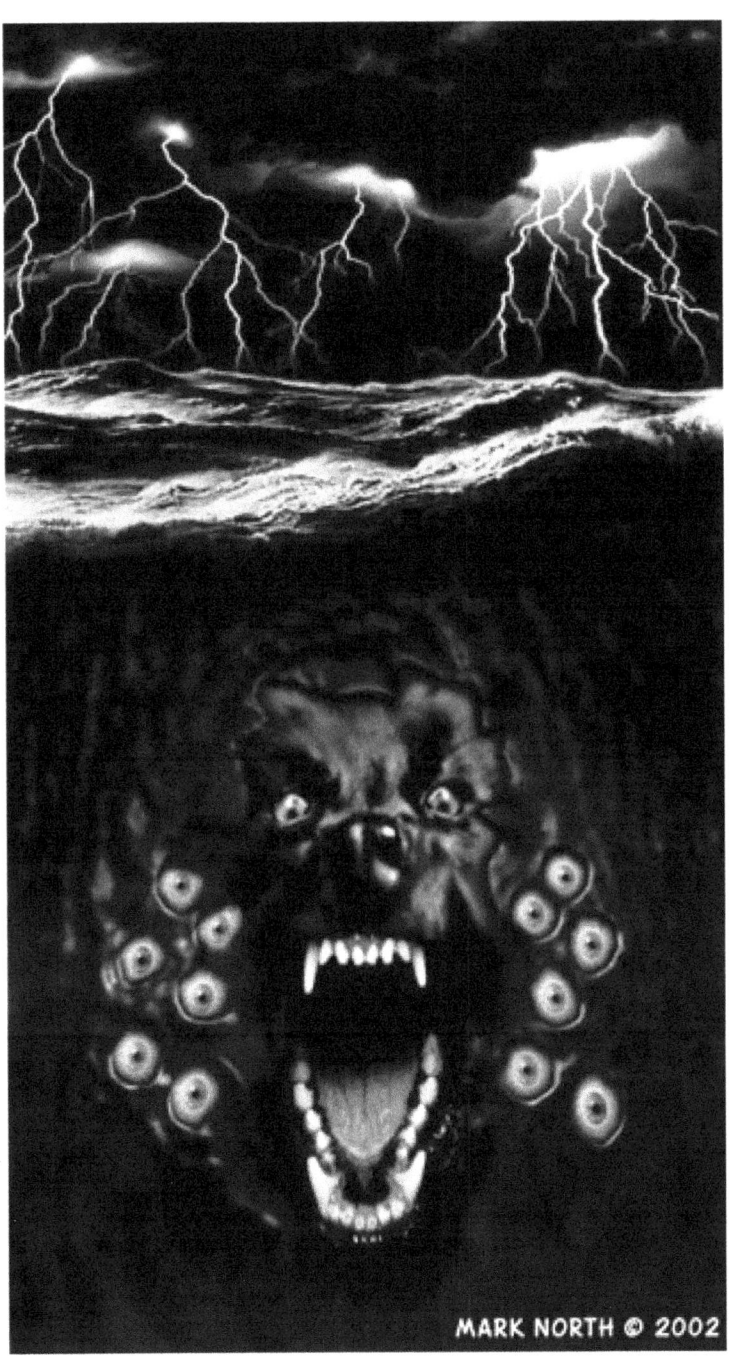

Beware the Roy Dog

hides in ponds, streams, rivers or lochs waiting to lure their victim to their watery demise.
In countries like France, we find that black dogs actually drown people, and in Belgium a goblin called the Kludde takes the form of a winged black dog, and attacks his victims in a similar way. Thus the Roy Dog may have been invented as some kind of bogey beast as a warning to those who were venturing too close to blow holes where air and water are forced through the natural opening by rising tides. Alternatively, perhaps as a reminder of even older times when cavern or tunnels where regarded as the entrance to the infernal regions, where fierce looking dogs, like the three-headed dog Cerberus in Greek mythology, and Garmr in Norse mythology, guarded the gates of Hell.

Like the Roy Dog, some black dogs in other parts of the country have also appeared during bad weather and it has been suggested that electrical storms might facilitate their appearance. It is this connection with environmental phenomena, such as electromagnetic radiation and ley lines, that black dogs often have the tendency to appear near these potential sources of energy.

Water is just another one of these environmental sources that black dogs are commonly associated with: one can find rivers, streams, lakes or the sea located close to the area in which the black dog is said to have appeared.

In Dorset, an example of this can be found along the River Stour, where a black dog is said to haunt the bridge at Blandford, and at a place not far from the bridge called Bryanston woods.

As recently as the early 1990s, a retired teacher, who was walking her elderly Labrador dog up the road that runs through the woods to Bryanston School, witnessed this animal. Her dog started to act strangely and tried to hide behind her legs. She looked up and saw a large 'Alsatian-type' dog coming towards them. As the animal drew closer she saw it was very dark in colour and possessed no collar, so she kept hold of her dog fearing that it may attack. When it drew level, she noticed that it was walking above the level of the road. On closer examination the woman described its coat as 'sticking out like spikes', and that its outline was fuzzy and indistinct.

Not far from the river is the village of Belchalwell. It was thought dangerous to go past certain gates at night because a black dog lurked there. This legend was supported when, during the Second World War, two men were driving along the road in the daytime, back into the village of Belchalwell, when they encountered a large black shaggy-coated dog, which jumped from the hedgerow on to the bonnet of the car. The weight of the creature actually caused the car to 'dip down'. The animal glared in at the two terrified young men for a few seconds, before it leapt up and floated over the hedge on the opposite side of the road. Although they both clearly saw the dog, describing it as 'very large and very, very black', no sound was heard and the car moved as if in a physical encounter.

I have also discovered that the Castle Inn at Durweston, another village that is close to The River Stour, was formerly called the Black Dog in the mid-19th Century, until its name was changed. One wonders if the pub was named after a forgotten local legend, or is it just another uncanny coincidence?

Pubs and Canine Spirits

In Dorset, black dogs can be found to be as restless ghosts that haunt particular areas where they once lived. One famous example, which features in both Devon and Dorset folklore, is the legend concerning the Black Dog of Lyme.

The tale begins at Colway Manor near the town of Lyme Regis during the 17th Century. A lonely old man, who once owned the manor, had, as his only companion, a loyal black dog.

One night, as he retired for bed, thieves broke into the house and demanded from him his hidden valuables, but the man refused. The thieves became angry and kicked and punched the man until he was dead. The dog however, was abandoned at the foot of the stairs to pine for his master until he eventually died of starvation.

The manor was almost completely destroyed during the Civil War, and a farmhouse was built on the remaining part of the mansion. At that time, the farmhouse retained the large original fireplace and two large antique seats, which were fixed to either side of the alcove of the fireplace. It was to these seats, every evening that the new owner - a farmer - would relax. One evening his solace was interrupted by the arrival of an eerie black dog, which came to sit on the opposite seat to him. The farmer was at first uneasy, but after a time he became accustomed to his new companion's regular appearances.

Discussing his strange visitor with neighbours, he was advised constantly to get rid of the creature. The farmer, who didn't fancy the idea of confronting the animal, jokingly replied, "Why should I? He is the quietest and frugalest creature about the farm, neither eating, drinking, nor interfering with anyone."

One evening while drinking with neighbours, the subject of his companion was discussed again. The farmer, who at the time was heavily drunk, got so fed up with their mockery that he stormed off back home to confront the spectral beast.

On his return, and in a terrible state of rage, he found the dog sitting at its usual place upon the chimney seat. The farmer, without any hesitation, seized a poker and lunged at the dog. The dog quickly jumped off the seat and fled upstairs followed in hot pursuit by the angry farmer. He soon cornered the animal in the attic, but the dog leapt through the ceiling and disappeared. Infuriated, the farmer struck a hard blow to the ceiling dislodging some of the plaster. From the hole, an old box fell to the floor. The farmer picked up the box to discover that it contained a considerable amount of gold and silver coins from the reign of Charles I. Could it be that this box contained the old man's valuables that he had concealed from the thieves that broke in that night all those years ago?

The farmer later decided to buy a house a mile west of Colway Manor, and with the help of his new found fortune, converted it into a coaching inn, where - in honour of his fortuitous companion - named it 'The Black Dog.'

The original coaching inn situated on the Devon and Dorset border remained at Uplyme until it was eventually pulled down in 1916 and a new inn, retaining The Black Dog name was built in its place.

This building still remains at Uplyme, where it once had the reputation as the first pub in Devon until its closure by the brewery in the 1990s. The property was boarded up for a while until it was bought and turned it into a refurbished bed and breakfast guesthouse. Now known as the 'The Old Black Dog' it continues to be run as a bed and breakfast business, by Mr and Mrs Dench.

But the story does not end there, for when the dog ceased its haunting of the farmhouse, it took to haunting, at midnight, the lane adjacent to the inn known as Haye Lane, alias 'Dog Lane.'

One encounter with the creature occurred late one evening in 1856. The witnesses were a local couple and the woman, whose occupation was a nurse, described the incident as follows:

> "As I was returning to Lyme one night with my husband down Dog Lane, as we reached the middle of it, I saw an animal about the size of a dog meeting us. 'What's that?' I said to my husband. 'What?' he said, 'I see nothing.' I was so frightened I could say no more then, for the animal was within two or three yards of us, and had become as large a young calf, but had the appearance of a black, shaggy dog with fiery eyes, just like the description I had heard of the 'black dog'. He passed close by me, and made the air cold and dank as he passed along. Though I was afraid to speak, I could not help turning round look after him, and I saw him growing bigger and bigger as he went along, till he was as high as the trees by the roadside, and then seeming to swell into a large cloud, he vanished in the air. As soon as I could speak, I asked my husband to look at his watch, and it was five minutes past twelve. My husband said he saw nothing but a vapour or fog coming up from the sea."

This account focuses attention on the very nature of the black dog phenomenon as in the incident involving a harbinger told by the Bridport lady in 1915. One person could see the spectral creature while the other companion could not.

The last reported sighting of the black dog was in 1959, and was seen by a family whilst on holiday in the area of Lyme. After visiting The Black Dog Inn, the three tourists were walking down Dog Lane when it came floating out from a hedge and crossed to the other side.

Since the 1856 sighting, the legend has been elaborated from its original version like so many other folk tales. Once being a harmless ghost dog of Colway Manor, it has now become a harbinger of death.

It has been suggested that if anyone encounters this apparition then death is imminent. However, if the victim can retain enough presence of mind to toss a silver coin to the dog, it will disappear and the spell of ill-fortune will be broken. A similar method was also applied to killing a shapeshifting witch hare, but as there has been no recent reported encounters with this creature there is no way of knowing whether the notion works!

Top: The original Black Dog Inn, Uplyme, before it was demolished.
Below: As it appears today.

Such elaboration of ghost stories like the black dog of Lyme, with its unfortunate new role as a harbinger, is said to also accompany another famous ghost of the town - the notorious and infamous Judge Jeffreys - who hanged and tortured many Monmouth rebels.

Another popular addition to the Black Dog of Lyme myth suggests that domestic dogs should, on no account, be allowed to stray late at night in this neighbourhood, as there have been many cases of their disappearance in a mysterious manner. This I jested with my colleague when visiting the area in 1998, when a poster advertising a lost pet border collie dog was pinned to a sign post at the entrance to Haye Lane.

The Black Dog of Lyme is just one of many recorded incidents with these spectral creatures haunting stretches of roads, country lanes, archaeological sites, and the parish and county boundaries of Dorset.

Around Hod Hill lurks the spirit of a black dog that was accidentally killed by a horse and cart after escaping from its cruel master. A broken chain still tethered around its neck can be heard as it runs through Stourpaine Village Square to Hod Hill. While another phantom black dog that continually haunts a quiet stretch of road was seen at Pot Lane in Horton near Cranborne.

One winter's night during the 1930s, Mr W. Armstrong and his companion were walking homeward to his cottage at Chideock along a moonlit stretch of road from Morcombelake. They were suddenly startled by the appearance of a large black hound, which followed behind them until they reached the cemetery north of Chideock. The two men stood transfixed upon the dark creature as it made its way to one of the large weathered tombstones, and to their amazement the dog simply vanished into thin air.

At Shipton Gorge, near Bridport, a man walking home late one evening was so startled by the appearance of a large black dog, that he picked up a stick and threw it at the animal, but to his amazement the stick passed straight through the dog, which then vanished.

At West Woodyates Manor, north of Sixpenny Handley, a similar incident happened to a farmer while unloading hay from his cart. He noticed, laying in the cart, a black dog, so with pitchfork in hand he struck at the animal whereupon it promptly disappeared.

Stranger still is the appearance of a headless black dog that, at midnight, is said to cross Bradford Lane at Dyke Head, a junction south of Leweston, near Longburton. A dog that is completely invisible at Pimperne can be heard rattling his chain whilst running along the Salisbury road from the foot of Letton Hill. Local people who have experienced the presence of this invisible hound talk of a soft velvety coat as it brushes past.

YO-HO-HO! AND A TIN OF CHUM

So, what may have contributed to these black dog legends to the county of Dorset? People who have owned a dog do not need to be told that they are the most obedient and affectionate animal companions, but for one to encounter a strange vicious dog can be one of the most terrifying experiences to be had. It is this fear that generates superstitions in isolated rural communities, and the development of tradition of folk tales and story telling.

But apart from fear, the black dog could bring some sort of reassurance to someone in need. For the people who have lost a family member or friends through strange and sudden illnesses, it could be a comfort to say that they died because the black dog had cursed them, and thus in their bizarre way, these phantom creatures could help explain the inexplicable.

Since the introduction of Christianity, the Devil was often the main contender for explaining the cause of any supernatural or unexplainable event, for it was the Devil's ability to change his appearance, so he could trick his unfortunate victims into a false sense of security before revealing his true identity. One identity the Devil often portrayed was a black dog.

One tale warning of the dangers of Sabbath-breaking happened in the early part of the 20th Century within the grounds of the old priory ruins at Woodcutts, close to the Wiltshire border. Young men from the village of Sixpenny Handley would often go there to play cards, but their afternoon game was interrupted one Sunday by the presence of a large black greyhound, with no ears and saucer-shaped eyes. It dashed across the room, disappearing through one of the priory walls. Some say that this creature may have been Satan himself in disguise.

We can say with certainty that a majority of black dog stories go back at least to the Elizabethan period, and that saucer-shaped eyes, which are often the traditional formulae to the black dog, were commonly attributed to 17th Century ghosts and demons. Obviously, 'tea-saucers' could not exist prior to the introduction of tea in the 17th Century. It is also this period in Dorset's history that we find the links with the Canadian province, Newfoundland.

The county of Dorset had been trading with Newfoundland from as early as the 16th Century when fisherman and merchants from Dorset began to exploit the rich supplies of cod fished off the coast of Newfoundland. As these trading interests grew more and more, people from the staple ports of Dorset went to work, and settled, in Newfoundland.

It is probably these fishermen, who worked off the coast of the eastern Canadian provinces, who first introduced both the Newfoundland and the St. John's Newfoundland (or the Labrador Retriever as it is known today) to Dorset and the rest of England. One tale to the first introduction the Black Dog or Newfoundland to this county can be found in Weymouth, which is uncanny as it was the first port of call of the arrival of the Black Death to this country back in 1348.

The first encounter with this breed is said to have taken place at one of Weymouth's oldest pubs. This pub, which had the reputation to attract the likes of smugglers, thieves and cutthroats, was originally called 'The Dove' until the 16th Century, when Weymouth won the contract to trade with the newly formed colonies of Newfoundland and Labrador.

The reason for the pub to have changed its name from The Dove to Black Dog, occurred when a sea captain from Newfoundland entered the establishment accompanied by what was described as 'a great black beast of a dog', and both the landlord and the locals where so amazed by what they beheld.

The Black Dog Inn Sign, Weymouth,
depicting the dove, the former name of the inn

The Portland Newfoundland

As the captain was retiring from his sea-faring days, he gave the dog as a gift to the landlord for his hospitality, thus introducing the first black dog to this part of the country. The local tale is that the dog brought such an amazing number of sightseers from the surrounding district, that the landlord changed the name of the pub in honour of the dog and the new found prosperity it had attracted for him.

Whether the Newfoundland black dog was a Labrador as depicted on the inn sign is questionable, as it could have well have been a Newfoundland breed, which would have made a dramatic impression upon the Weymouth folk at that time.

Whatever the case, it is a worthy tale to consider when researching the black dog legend in this area, as most inns called the Black Dog often depict the image of a Labrador as seen on the Weymouth pub. It is probable that other pubs near to the Dorset coast, such as the Black Dog Inn at Broadmayne and East Stoke near Wareham adopted the animal in the same way and, unlike the Black Dog of Lyme, there appears to be no association with a spectral black dog.

The Newfoundland also has the traditional formulae for the spectral black dog, being extremely large with a black shaggy coat and, lastly, its association with water. Charlotte Bronte makes this comparison in her novel *Jane Eyre*, chapter 12, with the meeting with Mr Rochester's dog, Pilot (also a Newfoundland) with Gytrash, Yorkshire's black dog. Even more uncanny, and more of a coincidence with reference to the black dog of Lyme, is that of the dog seen by Jane Eyre on a stretch of track called Hay Lane!

Both the Labrador and Newfoundland where versatile working dogs, able to rescue drifting nets and bring back shot waterfowl.

However, the Newfoundland was the much larger and stronger of the retrievers, capable of rescuing a drowning man or breaking the ice as he dove into the frigid northern ocean. The dog's lung capacity allowed the dog to swim great distances and fight ocean currents. At the end of a day's fishing, the day's catch was loaded into a cart, and the dog was hitched up to haul the load into town.

The origin of this working breed of Newfoundland is disputed. Vikings and Basque fishermen visited Newfoundland as early as 1000 AD and wrote accounts of the natives working side by side with these retrieving dogs. The breed as we know it today was developed in England, while the island of Newfoundland nearly legislated the native breed to extinction in 1780.

Whilst Dorset was trading with Newfoundland, we also find that smuggling was rife in Dorset and other neighbouring counties. So it might not be too surprising to find such a dog as the Newfoundland or Labrador, noted for its black colour and its agility in the sea. An example which made the news earlier this year was a black Labrador named Todd, who accidentally fell overboard from his master's boat in to the Solent near the Isle of Wight, only to travel ten miles to the mainland across cold choppy waters and a busy channel.

It would have been most certainly an ideal animal to train and use for the illegal purposes of retrieving smuggled goods that were moored off the coast, or from plundering wrecked vessels. In fact, in Dorset, Portland had its own breed of Newfoundland peculiar to the island.

The Portland Newfoundland was noted for retrieving small barrels of contraband spirits hidden offshore by local smugglers. The breed has been extinct on the island since the 19th Century. Could this be another possible candidate for the Roy or Row Dog myth?

As well as retrieving contraband, Newfoundlands or Labrador dogs may have been used as "Dog Horns", as it is a known fact that further west on the Devon coast, homecoming sailors were used to identifying the farms and villages onshore by the characteristic sounds made by different dogs when they barked. It has been suggested that some dogs were kept specifically for their penetrating bark, which carried well out to sea and were even irritated on purpose to encourage them to keep up the noise when fog was about.

It is only logical to assume that at some points along the Devon and Dorset coast, dogs would have been activated by fellow smugglers to signal luggers safely to the expected rendezvous that night. Places known as smuggling haunts, which are known to have black dog legends associated, were at Boatswain's Coppice - which lies between the two parishes of Lulworth and Tyneham - and at Fleet, which provided the basis for J Meade Falkner's famous novel *Moonfleet*. Though a fictional story, it had its foundations on the illegal trade that flourished here.

It is simple to imagine how a lonely traveller would feel if stumbling on such an operation. To see a large black dog, a breed that he or she may not recognise, wandering alone along the coast, or to emerge from the sea at night while their masters hid out of sight, it can be understood how it could easily be mistaken for some sort of demon or spectre of doom.

It was this very principle of deception to terrify locals that the famous Dorset smuggler, Isaac Gulliver, would often employ.

Gulliver once owned a farm near the mysterious Eggardon Hill hillfort - a Neolithic settlement - which still has the reputation of being a place with weird and unexplainable happenings, and is reputed to be the haunt of the Moon Goddess, Diana, who leads a ghostly pack of hunting dogs over its summit collecting the souls of the dead.

Gulliver would often use the routes that ran inland from East and West Bexington, Swyre, Burton Bradstock and Shipton Gorge to his farm. It was these local stories of phantom black dogs and spectral coaches that Gulliver and other smugglers delighted to keep alive, embroider and spread, so that as few people as possible would be out on those routes during the hours of darkness.

Gulliver is known to have used large black hounds, possibly Newfoundlands or Labradors, for poaching, and retrieving contraband and he disguised the coaches and wagons that were used for smuggling so that they appeared ghostly.

Hence, so many tales were spread around the area in which he and his associates would operate their illegal midnight runs. Like the tale of the headless dog reputed to cross Bredy Lane between Cathole Copse and Cathole Barn at Burton Bradstock, and the appearance of a ghostly black dog jumping over a wall on the Swyre road dragging a rattling chain near Puncknowle. Then there is the phantom funeral procession with four headless pallbearers, which haunts a place called Gadger's Hole, near Shipton Gorge, at midnight.

Of course, smuggling cannot explain all encounters, though it is factor to consider when investigating sightings that have alleged to have been encountered around that period.

So what about the more recent sightings? One might never know what people are seeing, though it has been recently suggested that big cats may have played a part and been mistaken for black dogs.

This may be possible as, at Sturminster Newton, there was once a terrifying creature, which ran along a track parallel to the main road at a place called the Hollow. The Rev J.B Hirst, a local clergyman in 1965, knew of someone who spoke of it as a large black dog.

Yet, there is a much older record, which shows it to have been a large cat, which appears in a 1907 manuscript *Reminiscences of Sturminster Newton* by Robin Young. He recalls the tale many years ago, of a wild and savage monstrous cat, with eyes as big as tea saucers, which is said to haunt the top of Newton Hill beside the ruined castle at Sturminster Newton. Local people were so afraid of encountering this creature that they would take the low road just to avoid it.

It is quite possible that this story supports Chris Moiser's suggestion that sightings of big cats during the 19th Century onwards may be the result of escapes from travelling circuses and menageries, as cages that housed these creatures were so badly designed that travelling across country on poorly maintained tracks, or drove accidents involving the toppling of the trailers, would result in large cat - or any other creature for that matter - to escape in to the countryside. It is also true that many aristocratic families kept exotic animals in private menageries, for example in the 15th Century the Martyn family of Athelhampton Hall, near Dorchester, kept apes as pets.

If a big cat or any other, exotic creature was released or escaped from a poorly maintained facility from the family estate, it would not be surprising why local people may have mistaken it as a black dog or another form of zoomorphic phenomena haunting a particular isolated area, which has a legend or story associated with it. There is the story at Friar Waddon near Portesham, where a farm labourer was warned about a large black cat with penetrating fiery eyes and a luminous tail.

As black dogs are less frequently reported these days, it is here where we can see that the big cat has taken over the role of this spectral creature. A role that has continued to do so even to this day, in the form of big cat sightings.

Sources:

Harte, Jeremy. *Cuckoo Pounds and Singing Barrows*. 1986
Mc Ewan, Graham. J: *Mystery Animals of Britain and Ireland*. 1986
Moiser, Chris. *Mystery Cats of Devon and Cornwall*. 2001
Morley Geoffrey: *Smuggling in Hampshire and Dorset 1700-1850*. 1983
Newland Robert.J. and North M. J:Dark Dorset: *Tales of Mystery, Wonder and Terror*. 2002
O'Donnell, Catherine via email (Black Dog stories from the Blandford area) 2002
Udal, John. Symonds: Dorsetshire Folklore. 1922
Sherwood. Dr. Simon *Apparitions of Black Dogs* website
Waring, Edward: *Ghosts and Legends of the Dorset Countryside*. 1977

If You Go Down To The Woods Today :
An Overview of Unsung Native American Cryptids

by

Mike Hallowell

Bigfoot – arentchasickovit? Well, no; but methinks that our pungent, hirsute hominid has been given a position of undue prominence within the field of North American cryptozoology.

In some respects this is understandable, of course. Of all cryptids within the Land of the Brave and the Home of the Free - or is it the other way round, I can never remember – Sasquatch is the most likely to be a *bona fide*, flesh-and-blood creature. Sasquatch is the current megastar of cryptozoology. If Sasquatch were a chef, Jamie Oliver would be relegated to working in a burger bar serving up lukewarm fries and strange things in baps.

If we are ever to catch a cryptid, then Sasquatch is the most likely candidate. This is why, like the '49ers during the gold rush, Bigfoot hunters are usually in a constant state of excitement. The Big Discovery just *has* to be around the next corner.

But there is a problem here. Cryptozoology has never been the exclusive domain of creatures, which fit comfortably into our system of taxonomic classification. Cryptozoology is also – or damn well should be – the province of much more ethereal creatures of a far more exotic nature.

Sadly – and this is no fault of Sasquatch hunters – the prominence of the Big Fella in the cryptozoological world has forced other, more stranger creatures into the metaphorical backwaters. If you think that Mothman is both the embodiment and consummate representation of the Daliesque cryptid, then you ain't seen nothin' yet.

Walk This Way...

There is an old Amerindian proverb, which states that you should never judge a man until you have walked at least three miles in his moccasins. In a broader sense, we can also say that we can never really understand a cultural perspective on anything until we have lived and breathed that culture.

Having a degree in Medieval History does not enable one to deftly wield a broadsword or hit a fly on a barn door with an arrow fired from a longbow.

I have Native American ancestry on my father's side (and allegedly my mother's side, although I haven't been able to substantiate this yet), and practice Amerindian spirituality. This has proved immensely useful when investigating cryptozoological mysteries, which have a Native American origin. When I first immersed myself in this field of endeavour, I realised that, from a non-Native American perspective, I was taking some tentative footsteps into what was, essentially, virgin territory.

Sasquatch is not a cryptid that has its origin in White American lore. Sasquatch was around before *wasichus* (a Native American term for white people) even knew that North America existed. But Sasquatch is only one of myriad cryptids found in Amerindian folk tales. Before we look at one or two, we need to understand why.

Picture This

The written word never played a part in Amerindian culture before the arrival of non-indigenous peoples. Everything of importance was passed down orally. When it was felt necessary to make a more permanent record, pictograms were used, although such occasions were relatively rare. Unable to appropriate hoards of clay tablets, papyri or stelae for study, historians and archaeologists were often nonplussed. The uncovering of vital Native American folklore was largely left, intriguingly, to a group of well-meaning amateurs who took it upon themselves to either learn native languages (and there are many) or find translators and interpreters who would facilitate the uncovering of tales that portrayed the essence of Amerindian culture. Two classics in the field are *Black Elk Speaks* by John G. Neihardt [University of Nebraska Press, 2000] and *North American Indians* by George Caitlin [Penguin, 1996].

Sadly, it has to be said that what has been saved is dwarfed by that which has been lost. The decimation of Native American tribes and the systematic destruction of their culture by *wasichus* have ensured that many Amerindian tales with a cryptozoological perspective have now been lost forever. Had European invaders taken a more enlightened approach, they could have uncovered and protected a cultural and spiritual heritage greater than that of Rome, Greece or the Renaissance. Instead, they chose to engage in a protracted period of cultural vandalism, which still has no parallel.

But there is a bright side. Native American culture and spirituality is breathing life anew. Many Native Americans are 'going back to the blanket' – that is, returning to their true heritage – and tales that were on the verge of extinction have been snatched from the jaws of oblivion. Hopefully we are entering a new era; an era in which serious cryptoozologists will have greater access to Amerindian tales. In these stories can be found a wealth of zooform animals – some semi-human – which may or may not have been real creatures in ancient times.

Talk To The Animals

Like no other people on earth, culturally and spiritually aware Native Americans have an intimate relationship with the flora and fauna of our planet. In some shamanic traditions, conversing with animals is no more likely to raise an eyebrow than conversing with your next-door neighbour. For this reason, animals – both taxonomically classified and cryptozoological – appear in staggering abundance in Amerindian tales. The sooner cryptozoologists begin to explore this virgin territory the better.

But there is yet another factor which needs to be taken into account. In the West, non-Native Americans believe that living things can be broadly classified into two divisions; animal life and plant life. Amerindians have never taken this view. They believe that everything in the universe is alive, including rocks, clouds, thunderbolts and raindrops. This may sound bizarre, but, after much study, I have come to embrace this view and believe that it has great scientific validity. Therefore, the notion of something *not* being alive – the concept of 'deadness' – does not exist in Native American tradition. This led Amerindians in ancient times to see interaction with the natural world, in all its aspects, as not just possible but almost inevitable. In many Native American tales, humans often intermarry with animals from other 'tribes'. These tribes may be conventionally recognised species, or cryptozoological beasts such as the Devil Fish which crops up in several Oneida tales.

But we can go even further. Interaction – and sometimes intermarriage – between different animal species and humans was also intertwined with a belief that humans could engage with *manitous* or spirits. There is an Algonquin tale, which tells of a powerful manitou that once visited the earth and fell in love with a beautiful woman. He married her, and she subsequently bore four sons. These were Michabo (the Friend of Man), Chibiabos (Lord of the Land of the Dead), Wabassa (the Rabbit Spirit) and Chokanipok (the Man of Flint). Here we see an emulsion of human, spirit, animal and mineral life forms. Vegetation such as trees plays an equally important part in other tales.

This extraordinarily flexible view of life and inter-species interaction allowed Amerindians in ancient times to incorporate a number of bizarre creatures into their folklore. After all, there are an infinite number of combinations to be had when every type of animal, plant, rock, spirit, gas, liquid and energy is thrown together in a melting pot and allowed to reproduce.

Let's Rock

In Iroquois tradition, there once existed a race called the Stone Giants who were violent, power-mad and without conscience. Having exhausted every weapon in their armoury, the five Iroquois nations (the Mohawks, the Oneidas, the Onondagas, the Cayugas and the Senecas) prayed to Tahahiawagon (the Upholder of the Heavens) and asked for his help.

Before long the Stone Giants decided to stop messing about and annihilate the Iroquois altogether. One night they crept up on an Onondaga encampment and hid. At dawn they planned to massacre every person there, man, woman and child.

But they had figured without the intervention of Tahahiawagon. Angered at the actions of the tyrannical stone giants, the Upholder of the Heavens came down to earth and, with a giant chisel, started chipping away at a nearby mountain. The resultant avalanche killed all but one Stone Giant. The sole escapee, Ganusquah, ran all the way to the Allegheny Mountains and hid. He too eventually met his demise, many years after, but under far different circumstances.

Short Service

Many moons ago, when the Cherokee lived in Florida, their natural enemies were the Iroquois. One raiding party, fronted by a chief, found that they had strayed far into Cherokee territory and had left themselves a long and dangerous journey back home.

At some juncture the chief became ill. He struggled on for as long as he could, but his comrades – such as they were – eventually decided that they could not afford to carry their stricken partner any longer. They agreed to leave him behind to his fate, and told him so. Naturally the warrior wasn't too chuffed about this, but he accepted his comrades' decision with good grace.

The other braves eventually made it home, but denied any knowledge of the chief's fate. To admit that they simply left him to die in the swamps would make them objects of hatred.

But the chief didn't die. Not long after his compatriots had departed, a small canoe pulled close to the shore. In it sat three extremely small creatures of mysterious appearance. To their credit, they promised to tend to the warrior until he was fully recovered. This involved travelling to a salt-lick some miles away, where they caught several buffalo. These animals they duly dispatched, and the flesh, hide and other parts were used to feed, clothe and treat the sick Iroquois chief. Once fully fit, these kindly dwarves even escorted him safely back to his own people.

When the chief arrived home, the game was up for the so-called friends who had abandoned him. The warrior told his people what had really happened. He also told them about the strange, dwarf-like creatures who had saved his life. He even took a group of Iroquois to the salt-lick where the creatures had killed the buffalo, but no trace of the dwarves was ever found. However, they did find the bones of many animals and arrowheads from arrows used by the diminutive swamp dwellers.

Evidence – and the Lack of it.

There are curious parallels between the above two tales and modern cryptozoology.

Take the Stone Giants, for example. Once populous, they are reduced to a single survivor who escapes into the forest. Like Bigfoot, he remains elusive. Indeed, the only witness to his existence was a Native American brave who he had kept prisoner for many years. When the brave is eventually allowed to return to his own people, the last Stone Giant crumbled into dust. The only definitive evidence that the creature ever lived – his body – had disappeared forever.

The dwarf-like beings who rescued the Iroquois chief were likewise never found. There was evidence of their existence – evidence of sorts. Their arrowheads were found. In like manner circumstantial evidence has been found to substantiate the existence of Sasquatch; photographs, film footage, hair tufts, footprints, nesting areas, etc. But proof absolute is always lacking. Like the Iroquois, we are always left to wonder.

Did the dwarf-like creatures ever have an objective reality? I believe so. Circumstantial evidence exists that a tribe of pygmies may once have lived in the Florida swamps. On a more Fortean note, it must also be said that almost identical beings known as the *Alux* once lived in Central and Southern America. David Hatcher-Childress, in his book *Lost Cities of North and Central America* [Adventure Unlimited Press, 1992], describes how small dwellings (or the remains of them), with doorways less than three feet in height, are found at the sight of nearly every Mayan temple. Scholars believe them to be nothing more than religious shrines, but locals are adamant that they were real dwellings once inhabited by the mysterious Alux. Reports of encounters with the Alux are still received today, the last, to my knowledge, in 1995. Loren Coleman refers to the Alux as pygmies, implying that they are nothing more than a race of extremely short - but in other respects normal – human beings. Personally, I'm not convinced. Their presence close to the Mayan temples suggests that they were viewed with a certain degree of awe, although Mayan sculptured reliefs indicate that they were also kept as slaves.

Special classes of persons - with specific religious duties, and who were both revered and yet, in a paradoxical sense, reviled - were not unknown in other parts of the ancient world. To me, this indicates that the Alux were perhaps not seen merely as diminutive humans, but as something more special.

It is indeed possible – I would suggest even likely – that the Alux share some common heritage with the dwarves described in the tale above.

The Stone Giants are something else altogether. They cannot be fitted into any taxonomic classification used today, and must, by any standards, be either be labelled as entirely mythical or something different entirely.

It is tempting to classify the Stone Giants and similar Native American zooforms alongside the Giants of Ancient Greece. However, this would ignore the radically different spiritual and cultural backgrounds of these two peoples. So, if the Stone Giants were not fictitious, what could they have been?

At this juncture it would be wise to bear in mind that there are two theories regarding how Native American peoples came to be in North America. Firstly, there is the generally accepted notion that they migrated from Siberia across the Bering Strait when this waterway was temporarily frozen over. Some theorists propose one, two, three or even more migrations at different periods in time. A less popular theory is that Native Americans came from elsewhere, and that they actually migrated across the Bering Strait into Siberia. Whatever the truth, it is a fact that the Native American peoples share a common biological heritage with the ancient (and modern) Siberians. Both peoples are born with what is commonly known as the 'Mongoloid Spot'; a small, blue bruise at the base of the spine which disappears after a while. This common heritage is interesting when it comes to identifying the nature of the Stone Giants.

The Siberian peoples and the Native American peoples also share many things in common spiritually and culturally. This can be seen in the study of tulpas – thought-forms which begin as an abstract idea and which then, through psychic or spiritual power, can be made real. Even westerners can be trained to produce tulpas if they devote the necessary amount of time and effort to the exercise.

Native Americans do not share the concept of producing thought-forms in the same way as their Siberian cousins. In fact, they do not really see zooform creatures as tulpas at all, but simply creatures of other natures and from other dimensions, which have a true external reality. Nevertheless, there are common denominators; the bizarre appearance of the creatures concerned, their transient nature, their spiritual and psychic power, etc. The bottom line is that creatures such as the Stone Giants can only be either incredibly powerful tulpas or creatures from an entirely different dimension.
The tulpa idea is tempting, but must be rejected for two reasons. Firstly, there is no evidence that Native Americans ever engaged in classical thought-form production. For tulpa production to work, one must first understand and believe in the concept. To produce an entire race of Stone Giants in this manner is something that not even the Siberians and others who borrowed the technique from them seem to have developed.

Secondly, Native Americans have a beautiful and highly-developed spiritual worldview which explains the existence of zooform creatures without resorting to the notion of tulpa creation. I believe indeed that they are 'real' creatures from other dimensions which occasionally irrupt into our own world for purposes we can only guess at.

The virgin territory of Native American spirituality and culture opens up two exciting possibilities for cryptozoologists, then. It allows for the study of bizarre creatures which, even to this day, may visit our world and dazzle our senses. It also opens up a way of studying the origin and purpose of these beings, giving us a way of understanding the Owlman, the Mothman, the Hopkinsville demons and, dare I say it, the legendary Sasquatch. Simply trying to verify the existence of such entities without understanding their cultural and spiritual context is a futile exercise. The Native American spiritual and cultural heritage is a tool which we can use to unlock the box which contains answers to our deepest cryptozoological questions. The fact that the most intriguing zooforms we know of appear in North America is no coincidence.

Exciting times lie ahead, if we have the courage to create them.

Fortean Zoological aspects of the `Rupert the Bear` stories

by

Richard Muirhead

Fortean zoology has long had a place in the antics and adventures of Rupert the Bear and his chums. As if talking animals (such as Rupert, Ottoline (a female otter), Bill Badger, Pong Ping (a Chinese Pekinese), Algy (a pug) and Podgy (a pig) - all in the 2001 annual) weren't fortean enough, the adventures the animals have in each story are no less such. For the purposes of this study, I examined Rupert annuals from the late 1960s up to 2002. I have managed to find a collection of stories ranging from the cryptozoological and mythical (Rupert and a sea serpent and a merboy, 1972) to the downright weird (Rupert and flying reindeer, 1986.)

Before an investigation into the stories themselves, it is necessary to provide some biographical background on Rupert's creators. The original founder of Rupert was Mary Tourtel, who was born in 1874 and who died in 1948. She first created Rupert in 1920 as the Little Lost Bear. The first Rupert cartoon appeared in the *Daily Express* on November 8th 1920. Tourtel was already established as an illustrator of children's books. She was the wife of one of the sub-editors of the *Express*, and she continued to draw Rupert until 1935 when her eyesight began to fail.

In 1935, Alfred Edmeades Bestall took over from Tourtel in drawing and writing the Rupert stories. He was born in Mandalay, Burma on December 14th 1892, the son of Methodist missionaries, and died on January 15th 1986 at Wern nursing home in Wales. Bestall went to Rydal School in Colwyn Bay, Wales. He obtained a scholarship to Birmingham Central School of Art. In 1915 he volunteered for the army and spent much of that year in Flanders. After the war he spent some time at LCC Central School of Art. From 1922 he spent some time as a freelance illustrator, contributing to periodicals such as *Punch* and *Tatler*. In 1935, Bestall wrote and illustrated his first Rupert story, *Rupert, Algy and the Smugglers*. He was 42. He wrote, up to 1965, a total of at least 273 Rupert stories. He also had the idea of introducing paper folding into the annual. In 1985 he was awarded an MBE.

He spent much of his life in Beddgelert, Wales. Bestall's favourite story was *Rupert and the Unicorn,* which first appeared in the *Daily Express* in 1964. The main current artist for the Rupert stories today is John Harrold.

In the 1968 annual, a story appears entitled "Rupert and the Fire Bird." Rupert and his friend Pong Ping are looking for fossils in a quarry. Pong Ping finds an egg but Rupert borrows it, and takes it home, followed by birds who urge him to drop it. On returning the egg to Pong-Ping, the Peke subjects it to various tests, including heating it (What *is* he going to do with the egg? thinks Rupert.)

Pong-Ping marches off into the countryside and agrees to meet Rupert at a certain spot "tomorrow." The next day Rupert and Pong-Ping make a circle out of paper on the ground, into which an aircraft lands. The chums open a box, out of which emerges a jet-black dragon, who breathes on the egg. It cracks open to reveal a small gold bird, which eventually flies off with the Nutwood birds. Soon a giant eagle takes Pong-Ping to a castle in the sky, as does a giant stork with Rupert. They are briefly imprisoned, but are soon summoned to explain to the king bird how they hatched the fire bird's egg. They explain, and the king summons the giant eagle to take the two chums back to Nutwood. Pong-Ping sends the dragon back to China.

The fortean zoological phenomenon of cats with "wings" is quite well known in fortean circles. In the Rupert annual for 1970 just such a creature appears in a story entitled "*Rupert and the Blunderpuss*." Rupert's mum is going on holiday and Rupert is left to roam. He spies two bat-like creatures. Later he sees flashing lights and experiences something like an earthquake. Rupert returns home where his uncle is looking after him, and his uncle sees a black cat with wings at the window. (*Upon the sill! A cat with wings! Oh my! I must be seeing things!*) Rupert goes out into the countryside around Nutwood and finds the black winged cat, who seems to want to play with him. Rupert then bumps into Tiger Lily, a Chinese girl, and together they find her father. He is a conjurer whose magic has gone wrong, thus creating the blunderpuss. Rupert goes off to look for the blunderpuss, to find it frightening his friends. He flatters it with kind and soothing words and soon the "cat" allows Rupert to pick him up. Rupert takes the "cat" home to show his uncle who says: "*I cannot see its loveliness*!" at which the "cat" takes offence and darts up the chimney. Rupert climbs up to the "cat" and retrieves it. He then returns it to the conjurer. The conjurer waves his wand and returns the winged cat back to the 'Land of Mystery,' upon which Rupert runs home to tell the tale to his friends.

The fortean aspect of the 1989 annual's relevant story, "*Rupert and the Boomerarrow*" is quite small but significant: it is the phenomenon of being carried off by a giant bird. Rupert and Bill Badger find an arrow stuck in a tree trunk. They free it and later bump into Bingo "the clever pup." Together they prepare to fire the arrow. Unfortunately, as Rupert is holding it, Bingo accidentally fires it and Rupert goes hurtling through the air. A huge eagle like bird catches Rupert, and the arrow, and takes Rupert to the castle of the Bird King. (Then, catching Rupert round the waist, it bears him off in angry haste.) The birds take offence at Rupert's intrusion and they put him in a cage. Not even a rescue mission by the Professor can help him. (Imprisoned on a windswept tower, Rupert grows glummer by the hour.) At some point the boomerarrow starts to shake, and then the arrow and the cage - with Rupert in it - flies through the air to be rescued by the Professor in his plane. They return to Nutwood where Rupert is reunited with Bill, Bingo and Rupert's parents.

In the 1997 annual there is a quasi-fortean tale about a stranded whale and Rupert. Rupert, his parents and Bill Badger are on holiday in Rocky Bay. Rupert and Bill climb up a small headland to meet Cap'n Binnacle. He tells them that he has seen a school of whales swimming in the bay. The next day Bill and Rupert go down to the bay where Rupert finds a solitary stranded whale. (*The mighty creature spots the pair and gives them both a mournful stare....*)

Then Rupert and Bill meet some sand sprites that attempt to help the whale by dousing it with water. Rupert and Cap'n Binnacle, however, decide to get help from King Neptune. They row out to sea and are met by a merboy, who takes them to an island where a giant turtle, the Tide Master, sits at a desk controlling tides. They persuade the turtle to arrange a high tide at Rocky Bay. Then, Rupert and the Captain return to the stranded whale and place inflatable objects like rubber rings under it. When high tide comes Rupert, Bill, the captain and the sand sprites all combine to push the whale out to sea, where it joins its joyful companions. Finally, when the excitement is over, Rupert and Bill and the Captain leave the beach and are reunited with Rupert's parents.

Rupert annuals between 2000 and 2002 were looked at but no fortean zoological stories could be found.

The monsters of Rupert Bear

Richard Freeman

As a boy I watched the TV puppet series of Rupert, and one character still looms pre-eminently in my mind: Raggety. He was an evil woodland sprite - a kind of demonic stick man. His face resembled a withered turnip with the straggly root as his nose. His body was constructed of gnarled twigs like a marionette built by Ed Gein. The striped jersey he wore threw this into sharp relief. Raggety had an awful mocking voice and danced a furtive little jig. I found him genuinely disturbing as a kid. I think there might be something about him in *Revelations*.

A pet baby dragon usually accompanied Rupert's Pekinese chum Ping Pong. Rupert's writers did not do their research on this because he is always portrayed as a perfectly formed miniature version of the adult dragons that turn up in the stories. In China, newly hatched dragons (whose eggs resembled precious stones) looked like ordinary snakes. Hence the saying (popularised by the 1970s TV series *The Water Margin*) *"Do not despise the snake because he has no horns. Who is to say he will not, one day, become a dragon?"*

After 500 years, the snake had reached huge proportions and developed a head that resembled that of a carp. In this form it was known as a *kiao*. A further 500 years and the creature developed into a *kiao-lung* with a reptilian head and four legs. After a further half century the dragon grew horns and in this stage was known as a *lung*. Finally, after yet another five centuries, it grew wings. In this ultimate stage it became a *ying-lung*. Hence Ping Pong's baby dragon should have resembled a snake.

Dragons of the Orient are still seen today, as the spate of sightings in the summer of 2002 in northeastern China's Lake Tianchi demonstrates. Another dragon-haunted lake is Lake Wembu in Tibet. This massive, and ill-explored body of water, is home to a reptilian monster the size of a house. Furnished with green scales, a long neck, and savage teeth, the Lake Wembo dragon seems not to have heard that Oriental dragons are supposed to be milder than their western counterparts. The monster is said to have smashed fishermen's boats and eaten the unfortunate occupants. Japan, too, has several dragon-haunted lakes such as Lake Ikeda, and Lakes Kutcharo, Nishikawa-cho, and Yamagata. In Japan dragons are called tatsu.

Sea serpents regularly turn up in Rupert stories. They are all long necked reptilian beasts. They closely resemble some of the sea serpents reported on the western seaboard of Canada and the USA. Dr Edward Bousfield and Dr Paul LeBlond have postulated the "mega-serpent" theory that these are a relic species descended from some pre-historic marine reptiles. Most famous among these are the creature known as *Hiaschukaluk* to the costal Indians of the north west and as Cadborosaurus, or Caddy, to the modern white men, and the San Francisco Bay monster.

Though not agreeing with all they write on the subject, I believe Bousfield and LeBlond are on the right track and these creatures are titanic aquatic reptiles. The reptile theory went out of favour mainly due to Heuvelman's ideas of unknown marine mammals. However, those who have seen the creatures up close often report scales. The "mane" running along these creature's backs may well be elongate dorsal scales like those found on iguanas and that we know some dinosaurs possessed. It should be noted that in all the earliest recorded dragon lore, in Sumeria, Babylonia, etc, dragons are seen as aquatic beasts as they are in most Oriental lore. One witness, who saw a specimen off Stinson beach during the San Francisco flap in the mid 1980s, actually described it as a dragon.

Most of the beasts that turn up in Rupert stories seem to have an eastern flavour. The Chinese phoenix is quite distinct to the Arabian phoenix - that one off specimen who died and was resurrected by its own flames in the sun temple of Egypt. The *fung whang* as it is called seems to have been some bizarre kind of pheasant with ornate tail feathers. Pictures of it in ancient Chinese texts on ornithology match no known bird. Together with the dragon, the tortoise, and the unicorn it was one of the four sacred animals in ancient China. Its appearance was said to show that the monarch was equitable and the kingdom had moral principles. The *fung whang* seems not to have been associated with fire unlike the more familiar phoenix. The Chinese phoenix was believed to be inherently female and the counterpart to the male dragon (contradictory when we hear of dragons laying eggs!). Pictured together the dragon and the phoenix were a symbol of marriage.

Unicorns appear in Rupert stories. Though they are often shown along side oriental dragons they are depicted as the western unicorn, a white horse with a single spiralled horn. The Chinese unicorn or *ki-lin* (ki-rin in Japan) was a very different beast. It was horse-like, but with scaled skin. It had goat-like cloven hooves and a backward pointing horn. It was said to appear at the end of a just king's reign. The birth and death of the philosopher Confucius was marked by a *ki-lin* appearing. Pictures of the *ki-lin* were attached to the cots of newborn babies. Both the eastern and western unicorns may have had their genesis in a form of antelope called *Procamtoceras*. This now extinct creature had two horns that grew so closely together that in life they would have been covered by one sheath giving the appearance of a single horn.

The fourth sacred beast was a giant turtle or tortoise that supported the sky on his back. Chelonian shells were used for divination in ancient China. Hot irons were placed on the shell of a dead tortoise after its body was removed. The patterns of the cracks in the shell would be interpreted to answer questions.

Sightings of sea turtles far, far larger than any living species prompted Bernard Heuvelmans to postulate the existence of a titanic unknown turtle he dubbed 'father of all turtles'. In the 3rd Century AD, Claudius Aelianus wrote in *De Natura Animalium* that turtle shells from the Indian Ocean measuring twelve feet were used as roofs for huts.

Al Edrisi wrote in *Description of the World* (published in 1145) that 30 foot turtles were to be found west of Sri Lanka. Some have even suggested that it may be a surviving form of the Cretaceous *Archelon ischyros*. This is problematic as all turtles must haul ashore to lay eggs. Hundreds of giant turtles crawling up beaches to lay eggs would soon be noticed. In order for a species of giant turtle to exist we would have to suggest a lost continent for them to breed on!

The answer to this riddle may be found in the male turtles. These animals never come ashore after hatching and can attain great ages. The largest species, the leatherback, can grow to nine feet or more in length. Could freakishly large male specimens exist? These would never come on to land and if seen at sea (where size is hard to estimate) they may be exaggerated into ship-sized monsters.

Why Rupert's monsters should have such an eastern feel about them is anybody's guess. Perhaps the artists had visited the orient or had some kind of interest in the area. And speaking as someone who has travelled in the Far East, I cannot blame them for having such a fixation.

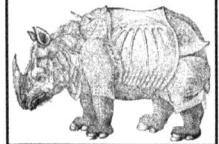

Cryptozoological reports from the Second World War

by Jonathan Downes and Nick Redfern

On the morning of 8 December 1941, the day after the unprovoked attack on the US Naval Base at Pearl Harbour, the Japanese invaded the British colony of Hong Kong. After bombing what was then the Kai Tak airfield, the Japanese crossed the Sham Chun River and advanced eastward towards the heart of the tiny British colony. After successfully occupying Tai Mo Shan, Hong Kong's highest mountain, the invaders fought their way inexorably towards Kowloon. By 11 December, the order to send defence troops to Hong Kong island was given and the last troops were ferried to the Hong Kong island on December 13. The only British Colonial possessions on mainland China had fallen in five days.

The next eight days saw a gallant and bloody battle for control of Hong Kong Island itself with two Brigades in charge of the defence of the Island. The East Brigade comprised the Rajputs, Royal Rifles and three companies of the H.K.V.D.C. The West Brigade comprised the Punjabis, Winnipeg Grenadiers, four companies of the H.K.V.D.C. and the Royal Scots. Each brigade was supported by artillery.

On 18 December, the Japanese started bombing the shoreline from Causeway Bay to Lei Yue Mun; and on that night, the Japanese landed at three points all in the Rajputs' sector between Lei Yue Mun and North Point. They immediately advanced to the high ground and captured

West Brigade headquarters at Wong Nei Chong Gap the next day. The defence there was cut in two when the Japanese advanced downhill to Repulse Bay on the southern shore. The East Brigade was forced back towards Stanley. On Christmas Day, on the advice of General Mattby, the Governor of Hong Kong announced the surrender.

The Japanese then embarked on a particularly brutal campaign of rape, murder and torture, and the entire European population was interned or deported to forced labour camps on the Asian Mainland. Most of the British civilian population was interned at Stanley Prison on the south side of Hong Kong Island, where they stayed until the liberation in the summer of 1945 - three and a half years later.

Amongst the internees was a young Biologist called Dr Geoffrey Herklots, who six years after the war was over, wrote a book called *The Hong Kong Countryside* which contained a remarkable story:

"During our internment at Stanley a remarkable story filtered into the camp that there was a tiger at large on Hong-Kong Island. Later it was reported to be on Stanley Peninsula. The guards got excited and it was risky walking about in the evening for an excited guard might fire at a prisoner mistaking him for a tiger! Soon pug marks were seen at the camp: I examined some myself but was by no means convinced. Then the story was spread that the tiger had been shot and finally there came into camp a Chinese or Japanese paper containing a photograph of the dead tiger. This photograph I saw. People said that it was a menagerie animal that had got loose; a likely story! It is strange how loath people are to believe that tigers do visit the Colony and occasionally swim the harbour and visit the island."

We are loath to appear judgmental, but it appears from the above passage that Herklots himself was not fully impressed with the truth of this episode. There are, however, several other pieces of supportive evidence, and it seems almost certain, again to paraphrase 'Alice', that 'someone killed something' and that the 'something' was a large tiger. The exact provenance of this tiger is less certain. '*Thagorus*' (1979) wrote:

"During the war, a tiger was shot by a party of Japanese Militiamen near Stanley in May 1942. A Mr E.W.Bradbury, who was once a butcher with the Dairy Farm Company, was brought from the Stanley Internment camp to skin the animal, the meat from which subsequently provided a feast for members of the Hong Kong race club. The animal was three feet high, six feet long, weighed 240lbs and had a nineteen inch tail. The skin of the tiger was stuffed and mounted in the hall of Government House, from which it was subsequently transferred to Japan in 1944.

"One theory about its presence on the island was that it had escaped from a menagerie during the Japanese invasion; another and more likely theory was that it had swum over from the mainland".

Although there are discrepancies between the two stories we shall avoid discussing them for the moment. Let us examine the supportive evidence for the claims. There is no doubt that a tiger is capable of swimming from the mainland to the island of Hong Kong. There is also no doubt that whereas the South Chinese race of the tiger *(Panthera tigris amoyensis)* is now highly endangered (there may be as few as eight wild specimens left at the time of writing), the quondam British Colony of Hong Kong was, until recently at least, well within the habitat of this beautiful and fearsome creature, and that Tigers have appeared in Hong Kong on a number of occasions - possibly as recently as 1977.

So, what is the problem? A third account of the 1942 Stanley Tiger, whilst initially appearing to be valuable and exciting evidence in favour of the event, actually casts some important doubts on its veracity. In 'Captive Years', their study of Hong Kong under Japanese occupation Birch and Cole (1982), describe conditions in Stanley Internment Camp, (now Stanley Prison).They quote a newspaper story from *'The Hong Kong News'*, an English language newspaper published by the occupying Japanese:

"Fierce Tiger shot in Stanley Woods!

"Successful Hong Kong police hunt in early morning.

"Although for some years past, rumours had circulated that there were tigers roaming the Hong Kong hills, it was only yesterday morning that such was shown to be fact and the feat of shooting the first tiger on the island was accomplished by Nipponese gendarmes and Indian and Chinese police at the back of Stanley village. Early yesterday morning the lowing of wild beasts was heard by many residents in Stanley village and gendarmes and police and military set off fully armed to search the hills. The search party consisting of Nipponese gendarmes and Indian and Chinese policemen was headed by Lt. Colonel Hirabayashi. The party was divided into smaller groups and a net was spread around the woods. After going over the ground for some considerable time, one group of searchers came across the tigers lair. They immediately opened fire but despite all efforts and the use of big wire netting the beast succeeded in evading the hunters. Not discouraged by the failure of the first attempt, the Nipponese police continued their search and a bigger cordon was thrown around the whole area.

"Apparently alarmed by the noise the tiger rushed about the forest for some time when it was again encountered by the police party. The police opened fire, and shots from an Indian policeman this time found their mark, causing the tiger to halt. The Indian fired three shots, hitting the tiger in the head, left shoulder and lungs".

Although not quoted by Birch and Cole, the article continued:

"Despite its wounds the tiger continued to struggle against the efforts of the policemen to tie it up. In the struggle, one of the Indian police was injured.

"However, the work of the Nipponese and Indian Police was fully rewarded as the tiger was finally subdued.

"The dead Tiger was then taken to the vacant ground outside the gendarme office at Stanley". This is, presumably, the newspaper article to which Herklots refers. Birch and Cole's book also includes a photograph of the dead beast, credited to 'Lady May Ride', which is captioned:

"The famous Stanley Tiger which was shot by the guards in 1942. This appears to be the only unofficial photograph taken by an inmate at Stanley".

The first thing that has to be ascertained is the identity of Lady May Ride. The only 'Ride', referred to in the text was Colonel Ride, the leader of the British Army Aid Group, the organisation which helped British Servicemen and other internees escape. Whether or not 'Lady May', is/was the wife of the gallant Colonel, it is unclear if she is the copyright owner or if she indeed was the photographer.

If this is the photograph from the Japanese newspaper, referred to by Herklots, why was it taken by an internee, whoever he, or she, was? Collaborating with the enemy to the extent of becoming an unofficial press photographer for a newspaper full of propaganda, which was published by the occupying power, would have been considered almost treasonable!

However, when we finally obtained a photocopy of the original newspaper cutting, we could see that the 'Lady May Ride' photograph was not the one used by the occupying Japanese. This is a far less impressive piece of photography than the 'Lady May Ride' photograph which, despite the claims that it is an 'unofficial' photograph, is obviously posed and well composed. The stringency of Japanese security arrangements, especially earlier on in the War is documented over again in this book. Violence, torture and even executions were relatively commonplace for what the Japanese considered to be infringements of security. If, indeed it was taken by an internee and not by a Japanese press photographer, then the evidence suggests that it was done so with the connivance, tacit or overt, of the Japanese Military. The head of the creature is being supported by a man who appears to be an Indian - presumably the policeman that shot it. If the man in the picture is a guard/policeman, as seems probable, he was certainly aware that he was being photographed. He is even smiling for the camera! It seems almost impossible that the Japanese Security Forces could not have been aware of the photograph.

When one compares it with the crude picture, which accompanied the item in *The Hong Kong News*, then the whole affair becomes even more unlikely.

Two seeming anomalies can be cleared up immediately. The guards/policemen are referred to as Indians and Formosans. Formosa (TAIWAN), was at the time a Japanese Colony. And although as Oliver Lindsay wrote:

"The Japanese put great pressure on the Indians to turn traitor against Britain. the vast majority remained loyal",

it implies that some, including the Indian man who is seen clutching the head of the Stanley Tiger, did not. Lindsay continues:

"The guards were later Formosan (Taiwanese) and were pettily officious and quick to take offence".

There is, however, another paradox. There may have been three thousand internees but it seems almost impossible that Herklots (who was after all Hong Kong's leading naturalist, the editor of the Hong Kong Naturalist magazine, and a minor celebrity in his own right), would not have known about the tiger incident from more than hearsay and rumours. Dr Herklots was important enough to be put in charge of revitalising the post war fishing industry for the region, in a successful attempt to restore food stocks as quickly as possible. Welsh (1993), gives more details of this affair and implies that Herklots, whom he describes (p.433) as a 'Biologist just released from Stanley Internment Camp' was a person of considerable importance. Even if he had not been taken to view the carcase in person, it seems certain that the photographer, who did see the carcass would have spoken to Herklots about it!

Herklots' testimony is beyond question. He was a reliable and indeed an expert witness; and whilst his mind may have been vague about minor details, surely an event as important to the sum total knowledge of the zoofauna of Hong Kong as this, would have remained fresh in his mind. As Forteans, the present authors are often accused of paranoid conspiracy theorising, but in this case, something doesn't add up!

The mounted skin was taken to a place of honour in the newly restored Government House and eventually to Tokyo as a trophy of war. The occupying army were inordinately proud of their trophy! At the time the Hong Kong News reported:

"A party of press-men, invited to Stanley to see the tiger yesterday morning found it weighed about 240 lbs and measured three foot high, 73 inches long with a tail of 90 inches (...) According to the Chinese, the appearance of a tiger is an omen of the approach of a period of prosperity".

It seems likely that the invading Japanese were determined to extract the maximum publicity from the event by exploiting local folk beliefs. Near the end of the war when it was obvious that they would lose, the Japanese were still fermenting Chinese Nationalist feelings, often through the use of cultural motifs, and sometimes by recruiting collaborators, in an attempt to ensure that at least the British would no longer be in power in Hong Kong. They, as history has proven, failed, but what seems almost certain is that forty years later when Birch and Cole were researching the incident for their book, someone, either wittingly or unwittingly, was not telling the whole truth!

Herklots was not the only person to report rumours that the animal had in fact escaped from captivity. Writing in 1978 Lindsay said firmly that *"it had escaped from a circus during the invasion",* and had therefore only been on the loose for five months. It would be interesting to know whether he had any further evidence to support this supposition and was not just sharing in the view, so scorned by Herklots, that bona fide wild tigers never actually visited Hong Kong.

The original report in The Hong Kong News gives (albeit unconscious) support to this theory by reporting:
"Stanley Villagers state that a tigress and three cubs, and a leopard were seen to have evaded the search party yesterday morning"

and a week later the same newspaper reported:

"Hunt for Tiger proceeds:

"Although one tiger has already been shot, the hunt is still going on for the other beasts reported to have been seen at Stanley.

"On Sunday last, about 200 gendarme police went to Stanley early in the morning and saw in the vicinity of Stanley Fort a black animal resembling a wolf, as well as a tigress with three cubs running about. The hunt was immediately taken up, but though two pigs were used as bait, it proved to be unsuccessful".

Although both leopards and wolves have at various times in history been reported as part of the pantheon of Hong Kong zoofauna, the appearance of them all at one in the same place suggests either mass hysteria on the part of those in the area at the time, or that the occupying forces had liberated (either by accident or design) a number of exotic animals in the area. One is left to speculate that the 1942 Stanley Tiger may not be a genuine example of a rare animal visiting the Colony. If it was, in fact, an animal brought in from somewhere else and released so that it could be killed as a potent piece of psychological warfare, then the incident is something far more rare and far more interesting to the fortean and to the student of military history!

Although the term wasn't even coined at the time of the events recounted in this book, Cryptozoology is widely accepted by both scientists and Forteans as an exciting and valid discipline within the natural sciences. As American cryptozoologist Chad Arment has written:

"Cryptozoology is the search and study of animals which are only rumoured to exist. It is a combination of biology and sociology. If it was completely zoology, the discipline would be redundant. The discovery of new organisms is exciting, but is well within the boundaries of Zoology proper. The essence of cryptozoology is the prior observation of these animals by non-professionals, and its subsequent revelation through myths, legends, or newspaper articles."

Essentially this is a description with which we would agree, and it is a discipline, which one of us has decided to make into his life's work. The preceding story about the tiger shot at Stanley Prison Camp is doubtlessly fortean in its nature but it is not cryptozoological, because it deals with a well-known species of animal, albeit in an exciting new context.

However, with hindsight the Second World War was an exciting time for cryptozoologists *per se* because for possibly the first time in history, hitherto largely unexplored areas, particularly within the Pacific theatre of war, experienced an influx of armed men, alert for the presence of the enemy. It is not surprising, therefore, that the annals of crypto-investigative research contain a number of interesting sightings of presumably unknown creatures made by members of the armed forces.

As The Japanese Imperial Army inexorably advanced across southeast Asia, and over-ran the European colonies that were scattered across the region, it looked ever more likely that the island continent of Australia would be the next to be invaded.. By 1940, following a period of political unrest in Australia which culminated in an extremely acrimonious general election, a Japanese invasion was a very real possibility. The Japanese Air Force conducted dozens of air raids upon the northern parts of Australia, and the Australian armed forces, for the first time in their history, prepared to defend their homeland.

There were over a hundred air bases in Queensland alone, and military patrols combed the area for signs of the invading Japanese forces that everyone believed would be landing from the submarines which regularly patrolled Australian waters. The invasion never actually came, but - from a cryptozoological point of view - a number of interesting observations were made.

According to the notorious Australian fortean author Rex Gilroy:

"....in 1943, a unit of soldiers patrolling the jungle near Normanton on the Gulf of Carpentaria coast surprised a large black catlike animal engaged in tearing up a full-sized bull it had just killed. As the men grabbed for their rifles, the animal left its 'kill' and dashed across foliage toward the men. A lucky shot brought it down. The soldiers were miles from nowhere and had to leave the body where it lay. Despite a later search, they failed to relocate it."

If you look in any respectable book on native Australian animals published before the 1960s, you will find a description of an animal called the Queensland Marsupial Lion or Tiger. No specimen was ever recovered of this critter, but the number of white settlers who reported such a beast convinced the folks in charge that something was roaming the bush in the far north of Queensland. The original inhabitants of the area had names for a particularly vicious animal, which could climb trees; they called it 'Yarri' and it was an animal of which they were justly wary.

The existence of such a creature was supported by fossil remains as young as 10,000 years old found in Australia of a largish tree-living marsupial predator named *Thylacoleo carnifex*. 19th Century Palaeontologist Richard Owen described this creature as being *"... one of the fellest and most destructive of predatory beasts."*

The Queensland Marsupial Lion/Tiger/Cat is usually described as a heavy-set animal about the size of a large dog with stripes across its whole back. It has a feline head and a nasty temperament, often taking its temper out on dogs sent out at it. Numerous reports describe it leaping through the air and disembowelling dogs with a swipe of its claws. Many reports indicate that it is a marsupial, with a peculiar hopping gait, and capable of great leaps; and an investigation of carcasses has revealed a pouch in the female. The most likely candidate for the identity of this animal is either a surviving population of *Thylacoleo* or a descendent of these creatures.

From a zoological point of view Australia, is a particularly strange place, because its animals are almost entirely unique. Although Australia was once connected to mainland Asia, the land bridge ceased to exist about 64,000,000 years ago. Apart from a few small rodents (brought by man) and bats (which flew), and some long extinct, and little known animals, like Tingamarra, a tree-shrew like creature from the Eocene which was found in fossil form at Riversleigh in about 1992, all the mammals in Australia are Marsupials, or the even more primitive egg-laying Monotremes. Man is a relatively recent arrival, having colonised the virgin continent in crude boats across the sea from what is now Indonesia.

Through the marvel of evolutionary divergence, marsupial mammals have evolved to fill every ecological niche on the continent. The grazing animals, such as deer and antelope, of Africa, Asia, America and Europe have been replaced by kangaroos and wallabies. The ecological niche left unfilled by the larger carnivores like wolves and big cats has been filled by animals such as the Tasmanian Wolf, and the supposedly recently extinct (in palaeontological terms at least) Thylacoleo. If these animals have, indeed, survived into the present day, it would be a zoological discovery of momentous proportions.

Most of the early reports of the Marsupial Lion centred around northern Queensland with its dense and sometimes inaccessible tropical rainforest. It was this very area, which, following the Japanese air raids of the early 1940s, was, for the first time regularly patrolled by soldiers. In the 1940s another flap of sightings of a striped tiger-like beast was reported further to south around Maryborough and Gympie, which lie just to the north of the Sunshine Coast.

The 1943 incident from Normanton is particularly frustrating because as Gilroy went on to write:

"Only a carcass, skeleton, skull or captured living specimen of one of these carnivores will ever convince understandably sceptical scientists of their existence; but many, at least secretly, are convinced from the mass of sightings claims Australia-wide that these mystery 'cats' exist."

Gilroy described other incidents from the same period when mysterious animals were shot, presumably as a direct result of the political situation of the time which would doubtlessly have inspired even the most mild mannered farmer to behave in an uncharacteristically trigger happy manner.

For example:

"It was in July 1943 that a Hughenden farmer shot and killed a six-foot-long 'panther' on his property after it had been making repeated attacks upon his cattle. He buried the corpse in the bush near his farmhouse, but the location is unknown."

The renowned Australian cryptozoologists Tony Healy and Paul Cropper, writing in *"Out of the Shadows - Mystery Animals of Australia"* (Ironbark, Australia, 1994), reported another fascinatingly vague account:

"....during the 1930s and 1940s a wave of sightings of what appeared to be Queensland 'tigers' occurred in the rugged mountains and dense subtropical forests surrounding the source of the Brisbane and Mary Rivers. This group of sightings gave rise to a local name for the creatures: the 'Yednia tiger'.

"In 1939 or 1940, two local timber cutters, Charles and Nigel Tutt, claimed a close encounter with the 'tiger' near Mt Stanley. Charles, who described it as being six feet [1.8 metres] long, grey with very dark vertical stripes and 'shaped like a light weight tiger'."

Our final report from the war years in Australia, comes from Rex Gilroy (again) who reported that:

"In 1940 outside Strathgordon, west of Lake Gordon in the wild bush country of south-western Tasmania, a farmer and his wife spotted a "seven-foot-length giant panther-like animal" as it dashed across their back paddock and leapt effortlessly over a six-foot-tall fence before disappearing into trees."

But Australia was far from being the only theatre of war where strange and mysterious creatures were being reported.

Every culture on Earth has its legends about man-beasts; shambling, hairy, humanoid figures who wander purposefully through the wilder parts of the countryside. The people of Nepal, Tibet and Northern India, believe unequivocally in the Yeti, the 'abominable snow-man' of the Himalayan Mountains. These creatures have been seen many times, well within living memory. However, there has never been a conclusive photograph, many of the most famous pictures of footprints in the snow have turned out either to be fakes or misidentifications of well known phenomena, and all the so-called yeti 'scalps' have turned out to be man made artefacts, cunningly fashioned from the pelts of local domestic livestock. Many scientists, however, still believe that 'the abominable snowman' is a living, flesh and blood creature.

In the late 1950s Bernard Heuvelmans, widely accepted as 'The Father of Cryptozoology', proposed a scientific name for the Yeti, calling it *Dinanthropoides nivalis*, - the giant ape of the snows. Several quite convincing theories have been promulgated to explain its identity. One theory is that all Yeti sightings are misinterpretations of encounters with known non-primates like the Tibetan Blue Bear - a rare sub-species of the Brown Bear. Another, is that the Yetis are nothing but wandering religious hermits, who avoid contact with other people to preserve their sense of spiritual harmony. The most popular theory with cryptozoologists, however, is that in parts of Asia there exists a relict population of a long extinct animal related to the Orang Utan. *Gigantopithecus blackii*, was first described after giant teeth were discovered in an apocathery's shop in Hong Kong by Dutch zoologist, Ralph Von Koenigswald, shortly before the Second World War.

After examination of the other fossil remains usually found along with those of *Gigantopithecus,* Von Koenigswald established that the giant ape lived some 500,000 years ago in caves in the Chinese province of Kwangsi, and could have lived as recently as 100,000 years ago, which is, after all, in purely palaeo-geological terms, a mere bagatelle.

It is the *Gigantopithecus blackii* identity for the Yeti and other enormous Asian apemen which seems to us, at least, to be the most likely candidate. However, the mystery remains unsolved; not surprisingly, however - in the light of the incidents from Australia that we examined earlier in this chapter - the military events of the first half of the 1940s produced at least two significant yeti sightings.

Of these, one was reported by a Polish refugee named Slavomir Rawicz. A Polish Army Cavalry officer, he was captured by the Red Army in 1939 during the German-Soviet partition of Poland and was sent to the Siberian Gulag along with other captive Poles, Finns, Ukrainians, Czechs, Greeks, and even a few English, French, and American unfortunates who had been caught up in the fighting. A year later, he and six comrades from various countries escaped from a labour camp in Yakutsk and made their way, on foot, thousands of miles south to British India, where Rawicz re-enlisted in the Polish army and fought against the Germans.

Bernard Heuvelmans takes up the story:

"They covered more than 3000 km on foot in country where nature was even more hostile than man. They crossed Mongolia, Sinkiang and Tibet, and in June 1942 the four survivors, indescribably thin and exhausted, were taken in by Indian soldiers.

"In May 1942 there were still five of them and they were near the frontier of Sikkim and Bhutan in the narrow Chumbi valley which leads from Tibet into the Indian plains. When they saw two black dots moving over the snow in the distance they hurried towards them, in the hope of catching some creature they could eat. But when they reached an overhanging rock about 100 m from the two creatures and 4 m above them, they saw their hopes were in vain."

The description given by Slavomir Rawicz is one of the best ever made of these fascinating and elusive beasts:

"They were massive creatures, nearly 2.50 m and standing erect. They were shuffling around on a snow-covered shelf which formed part of the obvious route for us to continue our descent. Armed as we were with only one knife and one axe the whole five of us together would have stood no chance against them.

"We thought perhaps they would go away if we waited. ft was obvious that the pair had seen us. They looked at us and were quite indifferent. They certainly were not frightened of us. I wanted to climb down the rock and get a little closer but my companions refused to join me. So we just sat there, our legs dangling over the edge of the rock, and watched them. For two hours we watched them. One of the party said that if we waited long enough we should be bound to see them descend on all fours. But they never did. They moved around quietly on their hind legs with an almost comical swinging action.

"As a regular officer in the Polish Army in 1939 was specially trained in assessing distances and heights for setting gun sights. I have not the slightest doubt that they were at least 2.10 m high, but probably nearer 2.50 m. One was slightly smaller than the other and we decided they must be male and female.

"Their faces I could not see in detail but the heads were squarish and the ears must lie close to the skull because there was no projection from the silhouette against the snow. The shoul-

ders sloped sharply down to a powerful chest and long arms, the wrists of which reached the knees. Seen in profile, the back of the head was a straight line from the crown into the shoulders, as somebody remarked at the time, 'a typical Prussian.'

"I tried to decide what they were and the nearest I could get was an idea of a cross between a big bear and an orang-utan type of ape.

"The nearest I can get to describing their colour is a rusty camel. They were covered with a long loose straight hair, which, in the light, seemed to have a greyish tinge, but the bodies seemed to be covered also with a very short reddish fur.

"They were doing nothing but move around slowly together and occasionally just standing and looking about them, like people admiring the view. So eventually we had to move off in another and more difficult direction."

This exciting encounter had tragic repercussions for the gallant refugees because soon after, one of the Poles slipped off a narrow ledge of rock and crashed to his untimely death. Slavomir Rawicz did manage to describe the 'half-blurred' footprints of these strange animals: they were apparently more or less oval in shape and were about 50 cm long by 20 cm wide at their widest point.

Writing in his seminal masterpiece *Sur la piste de betes ignorees* (which was eventually translated into English as *On the Track of Unknown Animals*) Bernard Heuvelmans describes another wartime Yeti sighting, again made by British soldiers stationed in northern India. Major Kirkland and Captain Maggs, together with a South African civilian called W.W.Wood, saw:

"...a large animal bounding towards us down the snow covered khud on the opposite side of the river. Its gait appeared to me to be that of a monkey in a hurry, with all four of its paws off the ground together, Maggs' recollection is that its rear legs were longer than its forelegs, and in running not very different from that of a rabbit..."

They described the animal as being the size of a man and tawny in colour. However, their reports also endowed this peculiar beast with a tail with bushy hairs at the end, which suggests that this was some kind of monkey rather than a *bona fide* sighting of the animal which, at the time, was generally known as 'The Abominable Snowman'. Unfortunately for those who like to see every anomalous encounter explained within the realms of accepted zoology, the largest known primate in the area is only thirty inches high, which leaves us with one of only two conclusions: Either, (as some commentators have suggested) the three witnesses were mistaken in their identification of the size of the creature, or they encountered something truly extraordinary.

During the battles between Colonial Dutch settlers and the invading Japanese in what is now Indonesia, two Dyak tribesmen encountered a strange man-sized creature. They later recounted their experience to the noted zoologist Lord Medway who passed on the information to the anthropologist and cryptozoologist Odette Tchernine. He wrote:

"Penyai saw the creature in 1941 on a height known as "Lost Hill", and they tried to kill it. The animal was considered half-beast, half-ghost. Its hair, said Betong, was black and dry, like palm fibres. Its footprints were gallon-sized. [sic] On all fours it was four feet high. On two feet it stood six feet tall. Betong also wrote: "This animal can go on all fours like a beast. It cannot walk upright like a man." This seemed a contradiction of his former statement indicating bipedal height.

"However, as Lord Medway wrote to me: "I think it is rather a good letter. You might care to send him a picture postcard of a gorilla!"

"And he gave me the Dyak's address which I have kept to this day, together with his letter in the Dyak language.

"Perhaps I shall chance Betong having moved and send him that postcard after all."

To some people, the possibility that we are not the only species of human presently living upon the planet, is an even more intriguing conundrum than the putative survival of a supposedly extinct ape from the Pleistocene era.

According to the Declaration on Race and Racial Prejudice adopted and proclaimed by the General Conference of the United Nations Educational, Scientific and Cultural Organisation at its twentieth session, on 27 November 1978: *"All human beings belong to a single species and are descended from a common stock. They are born equal in dignity and rights and all form an integral part of humanity"*. Whereas the ideology behind this statement is unquestionably sound, and it is not the intention of the present authors to question UNESCO or its stand against racial discrimination, we feel that there is a burgeoning amount of evidence to support the theories that. *Homo s. sapiens* is NOT the only species of man to live on Planet Earth.

Across what used to be Soviet Central Asia, as far west as what are technically parts of Europe and as far east as Mongolia, there are reports of hairy creatures known as Almas or Almasty that seem to be more akin to men than apes. Opinion is divided upon what these creatures are, or may be. It is tempting to theorise that they may be surviving pockets of our closest relative - the Neanderthal men - who supposedly died out in the later part of the Pleistocene epoch (more familiarly known as the Ice Age) some 200,000 to 30,000 years ago.

However, some researchers, most notably American anthropologist and cryptozoologist Professor Grover Krantz believe that they are true humans: surviving tribes of Mesolithic (stone age) hunter gatherers, similar to, but more primitive than, the aboriginal natives of the more obscure parts of South America and southeast Asia. Whichever hypothesis is true, the concept that primitive humans may have survived until the present day is a startling one, and one which has been bolstered greatly by a number of reports of these hairy men which were seen, captured and even shot during the war years.

Dmitri Bayanov, a Russian anthropologist, has collected a number of pieces of immensely valuable eyewitness testimony, including this report from Erjib Koshokoyev, of *Stary* Sherek in the Kabardino-Balkarian Republic - now a small Constituent of the Russian Federation which is situated in the Caucasus - that disputed area of far eastern Europe, which marks the eastern border with Asia and which has always been within the Russian sphere of influence from Tsarist times to the present day. Then it was a semi autonomous Soviet Republic and the scene of fierce battles between the Soviet and German armies.

We make no apologies for quoting Koshokoyev's testimony in full as it is particularly interesting:

"Before the war, there were many almastys around our area. Today, evidently, very few are encountered. I am somewhat informed about almastys, because I have heard a great deal spoken about them. I have personally seen them three times.

"The first time was in October, 1944. Our detachment of police was on horseback crossing a field of hemp, on the steppe... Suddenly, the horse of the first man stopped so abruptly that I almost ran into it: I was riding second in line... He said to me: 'Look!' An almasty just in front of us, a few metres away, an almasty was stuffing into its mouth the ends of stalks of hemp, with the grains on them. Behind us, the detachment was gathering around and making some noise. It saw us and ran away very rapidly - it ran extraordinarily fast - towards a shepherd's cabin which was not far away. While it was run-fling several men of our detachment took their rifles from their shoulders and prepared to fire, but our chief, a Russian officer from Naichik, cried out: 'Don't shoot, don't shoot! Let's capture it alive and take it to Nalchik.'

"We dismounted and surrounded the shepherd's cabin. We were quite numerous, and were able to form a solid circle around the cabin. I was just opposite the door, and saw every-thing very well. When we approached, the almasty came out of the cabin two or three times, in one bound. It appeared very agitated; it came out, moved around, jumped to one side, but then saw the men. It went back in one jump, immediately leaping out again, jumped to another side, but there also it saw the men. In doing this, it grimaced, with its lips moving very, very fast, and it mumbled something.

"Meanwhile, our cordon was approaching. We had closed ranks, and were advancing elbow to elbow. At this moment the almasty appeared again, jumped in all directions and, suddenly, gave a terrible cry and ran straight at the men. It ran faster than a horse. To tell the truth, the men were taken by surprise. It easily broke through our cordon, jumped into the ravine and disappeared in the brush surrounding the river.

"It was about 1.80 metres in height, and very robust. One could not see its face well because of the hair. Its breasts hung down to its middle. It was covered with long shaggy red hair, like that of the buffalo. The hair could be seen clearly through the pieces of the old handmade Kabardian caftan which it was wearing, and which was completely in tatters. "

Koshokoyev's account seems to imply that he considered this Almasty to be a man rather than a beast, and other testimony from nearby geographical areas would seem to confirm this.

One of the most controversial figures in both cryptozoology and forteana is undoubtedly the late Ivan T Sanderson. He has as many followers as he has detractors and this is neither the time nor the place to join in that particular debate. What is unquestionable, however, is that he amassed a great deal of anomalous data on the subject of strange animals, including another account from one of the Soviet Asian republics - again in the grip of a struggle against the invading Germans.

"From October to December of 1941 our infantry battalion was stationed some thirty kilometres from the town of Bumaksk (in the Dagestan A.S.S.R.) One day the representatives of the local authorities asked me to examine a man caught in the surrounding mountains and brought to the district centre. My medical advice was needed to establish whether or not this curious creature was a disguised spy.

"I entered a shed with two members of the local authorities. When I asked why I had to examine the man in a cold shed and not in a warm room, I was told that the prisoner could not be kept in a warm room. He had sweated in the house so profusely that they had to keep him in the shed.

"I can still see the creature as it stood before me, a male, naked and bare-footed. And it was doubtlessly a man, because its entire shape was human. The chest, back and shoulders, however, were covered with shaggy hair of a dark brown color. This fur of his was much like that of a bear, and two to three centimetres long. The fur was thinner and softer below the chest. His wrists were crude and sparsely covered with hair. The palms of his hands and the soles of his feet were free of hair. But the hair on his head reached to his shoulders, partly covering his forehead. The hair on his head, moreover, felt very rough to the hand. He had no beard or moustache, though his face was completely covered with a light growth of hair. The hair around his mouth was also short and sparse.

"The man stood absolutely straight with his arms hanging and his height was above the average about 180 centimetres (about 70 inches]. He stood before me like a giant, his mighty chest thrust forward. His fingers were thick, strong, and exceptionally large. On the whole, he was considerably bigger than any of the local inhabitants.

"His eyes told me nothing. They were dull and empty - the eyes of an animal. And he seemed to me like an animal and nothing more. As I learned, he had accepted no food or drink since he was caught.

"He had asked for nothing and said nothing. When kept in a warm room he sweated profusely. While I was there, some water and then some bread were brought up to his mouth; and someone offered him a hand, but there was no reaction. I gave the verbal conclusion that this was no disguised person, but a wildman of some kind. Then I returned to my unit and never heard of him again."

The Second World War broke out rather unexpectedly in the Soviet Union on 22 June, 1941. On the eve of what came to be known as "The Great Patriotic War," Baku Azerbaijan was the cradle of the Soviet oil industry, and as such, the major supplier of oil and oil products. In 1940, for example, 22.2 million tons of oil were extracted from Baku which comprised nearly 72% of all the oil extracted in the entire USSR. Consequently, the war could barely have been won had it not been for Baku oil and the fine quality of fuel that this city continuously supplied to the war front between 1941-45.

Hitler was determined to capture Baku's oil, and thus disrupt the major European source of petrochemical fuels, Even the date of attack was scheduled (25 September 1942). Anticipating a forthcoming victory, Hitler's generals presented him with a cake, fashioned to resemble Baku and the Caspian Sea. Delighted, Hitler chose the best piece for himself - Baku. Fortunately, the attack never occurred, and German forces were defeated before they ever reached Azerbaijan.

However, the advancing German war machine reached Dagestan by the end of 1941, and one can easily picture the level of paranoia that the relatively primitive Soviet Army must have felt, faced as it was with the prospect of an imminent encounter with the largest and most fearsome body of troops ever assembled. In the light of this, the scenario of a wild-man being arrested by the Germans on suspicion of spying - bizarrely reminiscent of the (probably apocryphal) story of the shipwrecked monkey which was hung in Hartlepool as a French spy during the Napoleonic wars - does not seem so bizarre as might otherwise be imagined.

Simon Welfare and John Fairley report another interesting series of sightings from the wartime period:

"In 1944 a hairy man was shot and killed in Tashkurghan near where the borders of India, China, Pakistan, Afghanistan and the Soviet Union all meet. Professor B A Fedorovich cross-questioned many of the witnesses subsequently and was convinced the story was true. A Moscow factory chief, G N Kolpachnikov, relates that he was leading a reconnaissance unit in Mongolia during the Japanese invasion of 1937. One night two sentries saw a pair of silhouettes descending the side of a hill. Challenged, they made no response, so the sentries shot them. Kolpachnikov describes his astonishment when he saw the bodies in the morning. 'They were not enemies, but strange hairy creatures more like an anthropoid ape. But I knew that there were no anthropoid apes in the Democratic Republic of Mongolia.' He questioned an old man who told him that wild men were sometimes encountered in the high mountains. Kolpachnikov recalls that the bodies were about the height of a man covered irregularly with reddish hair, sometimes thick but often with the skin showing through. The face was like a very coarse human face with prominent eyebrows.

"With a war going on there was no opportunity to do anything more than bury the bodies."

Such tantalising stories, are, unfortunately, a staple of cryptozoological research; and it is because of the lack of concrete information that creatures such as the reclusive Almasty - whether they be animals or men - still lurk in the socio-cultural grey areas between fact and belief. At least, on this occasion, there is a believable excuse for the non-recovery of the bodies.

We finish this chapter with a story that is very similar to the one with which we began. One of the most enduring contemporary animal mysteries are the alien big cats (ABCs) which have been reported over the last half century or so across much of the world - often in places where no indigenous species of large felid is to be found.

The so-called 'Beasts' of Exmoor and Bodmin are well known - those of Australia (such as the Emmaville Panther) less so, but it is back to Australia that we must go in order to find what may well be a convincing explanation for this enduring mystery. In 1994 Tony Healy and Paul Cropper wrote:

"Ever since the Grampians puma phenomenon came to the attention of journalists and cryptozoologists in the early 1970s, various local identities, such as the late Dick Saligari, have insisted they knew what the creatures were and where they came from.

"The animals, they said, really were pumas. They were descended from American mountain lion cubs which were brought to the Mt. Gambier area as US regimental mascots during World War II. When the cubs grew too large to handle they were supposedly taken to the Grampians and released."

As is so often the case, finding the source of such rumours is a laborious and highly complex task, but eventually they managed it:

"In the 1970s investigators found that while the US mascot theory was widely believed in the area, it was damnably difficult to get to the source of the story. Finally, in May 1989, an elderly Hamilton woman, Irene Addinsall, broke what she said was a self-imposed 46-year silence about the genesis of the big cats.

"Miss Addinsall, then 78, said she remembered seeing the regimental mascots several times when she was a Land Army girl on her uncle's property, 'Kangaroo Park', near Hotspur in 1943. An American unit to which they belonged camped on Crown land next to the property.

"There was a man among the soldiers with a light-coloured puma. She had four kittens, three light-coloured and a little dark-coloured one. They were always getting twiddled up in sticks or falling over. The army boss down there said he couldn't stand it. She was getting savage because the kittens were being hurt. She was becoming scatty. He told them to get rid of it!

"The boss went down to a party at Heywood and got a bit worse for wear. While he was away... they put the puma on a truck and took her up towards Halls Gap to one of those creeks... and they let her out there in the middle of the night. She didn't want to stay... wanted to come back with them. There were some rabbits and she ran after them and the kittens ran after her... that was the last they saw of her.

"A month after Miss Addinsall told her story a Maryborough man told the Advertiser he had seen a large cat with a chain around its neck in 1944 or 1945.

"He was a child at the time, playing beside the road when he heard the sound of a dragging chain. Looking up, he saw a jet-black felid the size of a large dog 10 metres away. The event supposedly occurred south of Halls Gap in the Grampians. If we choose to believe that the black 'puma' cub of Miss Addinsall's story was released with a chain still around its neck and

that it dragged the chain around as it grew up, then the latter story could, perhaps, be seen as corroboration of her account.

"Victorian big cat researcher Bernie Mace had no doubt Miss Addinsall's story was true. He later revealed that in 1987 an elderly man had told him, in confidence, that he had actually taken part in the dumping. 'He even mentioned Miss Addinsall by name ... it all fits'.

"It would certainly seem so. As alluded to earlier, some of the most interesting evidence uncovered by the Deakin University team in 1977 was found not in the field but in the archives. While attempting to nail down the regimental mascot theory, they eventually discovered the identity of two American units which were stationed around Mt. Gambier in 1942. They were the 35th and 46th Pursuit (Fighter) Groups. The units later moved to Queensland and then to the Pacific islands.

"Digging deeper, the researchers hit what appeared to be the jackpot: they found that the colour flash emblems of both groups displayed the outline of large black cats. Miss Addinsall's story, the eyewitness account of the dumping, the 'puma with the chain' story and the 'black puma' shoulder flashes seem, at first glance, to confirm the regimental mascot theory. As Bernie Mace said: 'It all fits'."

Bernie Mace visited the UK in 1990 when he met several researchers from the Centre for Fortean Zoology, telling them of this theory several years before Healy and Cropper eventually published their book *"Out Of the Shadows"*. Much to his amazement, the Devonshire based researchers told him that they had heard exactly the same story - this time from their own native county!

As mentioned briefly above the existence of mysterious big cats in the western counties of the United Kingdom has been a matter of scientific conjecture for many years now, and thousands of well attested sightings have been made. Even one of the present authors has gone on the record as having seen what appeared to be a female puma on Bodmin Moor during the summer of 1997.

It is popularly believed that these animals are the descendants of creatures liberated illegally in the wake of the 1976 Dangerous Wild Animals Act, which forbade the keeping of many species of exotic pets. However, although this is almost certainly true, it does not explain the sightings of such animals prior to the late 1970s. There are persistent rumours amongst fortean researchers that some animals - possibly Bobcats (North American lynx) were kept as mascots by American and Canadian troops stationed in the South Hams area of Devon during 1942-4, and were liberated there prior to the D Day landings.

Although this cannot ultimately be proven conclusively, it makes a certain amount of zoological and historical sense. If these animals were kept as mascots, what were the soldiers to do with them, once the invasion of mainland Europe was underway? With stringent food rationing for civilians, it is unfeasible that they would be left in England: They couldn't take them with them. The logistics involved in shipping the beasts back to the United States/Canada would be ridiculously complicated, and one would imagine that even the most hard hearted soldier would have been loath to euthanase the Regimental mascot.

Therefore, a surreptitious release would seem to have been the only logical solution.

Whereas the 1941 tiger only restored (albeit for a very short time) a once-indigenous animal to its native haunts, the activities of well meaning, but zoologically naive, soldiers in both Australia and the West of England, may have not only irretrievably altered the zoofauna of two completely separate ecosystems, but may also have unwittingly created two enduring and perplexing mysteries in the process!

Who would have thought that the mystery would still be raging half a century later?

Dr. Bernard Heuvelmans, often referred to as the 'Father of Cryptozoology', studied reports of anomalous marine creatures over a period of several hundred years, and after having rejected (a) creatures that were known to science; (b) sightings which he believed were too vague to classify; (c) obvious hoaxes, or sightings which he believed were inadvertent misidentifications of animate or inanimate objects; he hypothesised 9 basic types of sea serpents in his 1968 book, *In the Wake of Sea Serpents*:

1 - Merhorse - 40/100 ft long, large eyes, smooth skin and a mane
2 - Multi-Humped - 50/100 ft long, looks whale-like with several humps
3 - Long Necked - 30/70 ft long, small head and 4 flippers
4 - Multi-Finned - 50/70 ft long, whale-like, many fins, dorsal fin, armoured skin
5 - Super Otter - Giant otter-like animal, long tail, gray of beige in color
6 - Marine Saurian - alligator-like creature, long head and tail, scales
7 - Super Eel - 20/100 ft long, giant eel-like fish, no limbs
8 - Father of All Turtles - abnormally huge turtle
9 - Yellow Belly - 60/100 ft long, tadpole shaped, yellow with a black stripe

Unsurprisingly, several of the most important sightings of just such anomalous marine creatures were made by Naval vessels during the 1939-45 War! One particularly intriguing account is chronicled by Heuvelmans himself:

'Not all the eye-witness reports were of equal value, but if they were not detailed nor always convincing they were at least sincere. First there was the testimony of John Drummond, who was second officer on H.M.S. Buster of the Freetown Escort Force when he had a strange encounter off the Gold Coast at the beginning of 1944. The ship was patrolling at dusk, searching for a German submarine that had been reported. Drummond was on the bridge when he saw a big ray 5 to 6 feet in span, which despite its size leapt out of the water like a trout after a fly. He wondered what would have frightened it as to make it jump so high, when the lookout shouted, 'Huge object on port beam!'

'That [said Drummond] was gentle understatement. I have seen something of the like since - in horror movies, hut this thing, in slow motion, heaved itself out of the depths, remained suspended for four to five seconds and then fell forward with a thunderous splash. I think Able-Seaman Fitzgerald of Pimlico was the look-out concerned. I think that, because when I inter-

viewed him afterwards, he described the manifestation as - 'Like the side of a big building rumbling down in the London blitz.' The asdic operator on duty . .. held a sharp underwater contact on his set, then lost it on the recorder as the thing dived.

'Drummond immediately reported it in the log. He was convinced it had not been a big whale. This was at once confirmed by Lieutenant-Commander Oh Bernhardt Egjar of Tonsberg, a Norwegian, then attached to the Royal Naval Reserve who had hunted whales in Antarctica before the war.

'Then what was it? [wondered Drummond]. Distances at sea are often very deceptive. Our main mast was well over sixty feet high. The thing seemed about that height, at a guess. In the dusk I could see no eyes. The suggestion of a tail, yes. But the most vivid impression was the thinness of the creature.'

Although this anecdote makes thrilling reading, it is of little or no value from a scientific point of view. The creature was "about" sixty feet high and it had the "suggestion of a tail"; but the truth is that Drummond's account is so vague that we can not even be sure that what he saw was animate, let alone make any intelligent guesses as to what the animal (if indeed it *was* an animal) might have been.

Of somewhat greater value as a piece of anecdotal evidence is this account collected by author Michael Bright:

'...during the Second World War, in 1942 to be precise. Mr Welch was on board a troopship bound for Bombay from Durban. He was on look-out duty.

'We never knew quite what we were looking out for but we were always on the look-out. On one occasion, though, I could see a large black object way in the distance. My heart went down to my boots because I thought it was a submarine. I sounded the alarm, bells rang all over the ship, and everybody was going mad, panicking.

'One of the duty officers looked through his binoculars and said, 'Oh no, it's not a submarine, I don't know what it is, probably something just floating in the water.

'Anyway, as the ship got nearer we could see what I can only describe as a sea monster; it was definitely something swimming. It crossed our bows and we could see it quite clearly; it was a sort of a serpent, about 20 to 30 feet long, very thick - probably as thick as a tree-trunk, and its back was arched in several places.

'I couldn't make out its head with any clarity; it was sort of surrounded by waves. We carried on, and it went its way while we went ours; it took no notice of us whatsoever, and eventually it disappeared from view.'

Bernard Heuvelmans has suggested that these 'many humped' sea serpents are a surviving form of archaeocete - a very primitive whale dating from about 25 million years ago. Whales were originally land animals, and the most primitive example we know of was *Ambulocetus* -

a peculiar sealion-like creature known only from fragmentary fossils in parts of what is now Pakistan. As these creatures evolved they became more and more adapted for a completely marine existence. However, before evolving into the torpedo-like fish shape of all modern whales, some whales with a long, thin serpentine form did exist and it is tempting to speculate that the animal seen by Mr Welch in the Indian Ocean was one of these.

Some Sea Serpents are less easy to define. These are animals known to Heuvelmans as the 'long necked' and the 'merhorse' which he speculated were pinnipeds of some sort. Pinnipeds include seals, walruses and sea lions, and all known species have to come on land to breed. Although a recent fossil seal with an elongated neck has been found, there are no known species of long-necked seal alive today. A Dutch zoologist called Antoon Oudemans, who wrote at the end of the 19th century, was the first person to propound the theory that some sea-serpents are members of a hypothetical species of long necked seal; and whilst solid evidence remains elusive, many contemporary cryptozoologists believe that this theory is, indeed, correct.

One of the most famous sightings of what appears, from its behaviour and the description of a 'mane', at least, to be one of these hypothetical long-necked pinnipeds, took place during the War years and was first, or so we believe, collected by veteran Loch Ness Monster investigator, Tim Dinsdale:

'A man came in one day and asked to be introduced. Apparently he had been aboard HMS Galatea during the war as a Chief Petty Officer. It was November of 1940. The ship had been mined shortly before, and Mr. Ratcliffe, my informant, was badly shocked by the explosion. Whether this could have affected his judgement it was hard to say, but at sea once more Galatea a light cruiser, encountered an immense sea animal. A huge and ugly head, with a curious mane-like appendage draping down its neck extended upright from the sea, on a long neck-like protrusion. Behind, at intervals, enormous looping coil broke the surface.

'Mr Ratcliffe's dimensions seemed quite incredible and I could not accept them. However, I had no doubt about his sincerity.

'He told me than an order came down from the bridge of the Galatea for target practise with the forward 0.5 inch Machine Gun battery. This opened fire and Mr Ratcliffe who was on deck watched the tracer arc and saw the creature's head jerk and shiver as the bullets struck. Galatea was in action almost daily at the time in the North Sea and the English Channel which was alive with enemy aircraft. No doubt every opportunity at target practise was justified with men's lives at stake but when the creature disappeared from view the rumour quickly went round the ship that the ship was doomed for a certainty. the beast was not real at all but a manifestation of death, and to open fire on it was tantamount to suicide!

'In due course, HMS Galatea was sunk like so many other proud ships of the Royal Navy but it was not until December of 1941 that she rolled over to a torpedo off Alexandria in the Mediterranean.
'I later found this out from the Imperial War Museum; because Mr Ratcliffe very sensibly suggested that as all gun firing practises were logged I might find some mention of the monster

target recorded in the Galatea's log. The museum knew of no references to the incident and did not know if the log had survived; they referred me to another department. So there I let the matter rest.'

It is tempting to fantasise that the Galatea met her untimely end because of a jinx caused by her killing a mysterious sea-monster, but as rationalists we feel that this is highly unlikely. What is far *more* feasible is that the crew, already with a low morale due to the turn that the War had taken, and being part of the home defences themselves were shocked into even lower spirits by the events surrounding the unfortunate death of the sea serpent (a creature enshrouded with nautical myth and superstition). Indeed, their performance may well have suffered as a result and may have led to the unfortunate end which overcame the Galatea thirteen months later.

Coincidence? Maybe. However, there are other sightings of 'maned' sea-monsters from the war years:

As we have seen earlier in this chapter, the Gulf of Mexico was a hotbed of U-boat activity. When Mexico entered into the Second World War on the allied side it found its foreign policy had changed dramatically. Suddenly, Mexico found itself an ally of a country that had been its main enemy in the world. The US Government helped Mexico to get its first international for many years and also opened up its markets to Mexican goods.

To reciprocate, the Mexican government signed agreements concerning commerce, migrant farm workers, and military co-operation. Raw materials were sold to the US at lower than free-market prices. In exchange the Mexican government accumulated large reserves of US dollars that had to be saved - largely because there was nothing to spend them on. The US produced only for the war effort and Mexico was cut off from US imports. Thousands of migrant workers worked for US agricultural businesses, 15,000 joined the army, and 1,492 Mexicans lost their lives fighting for the US in the Pacific and Europe.

However, Mexico, as it is now, was a divided country with many links to Germany. Although we have not been able to discover any documentary proof, there is some reason to believe that some of the separatist landowners in the southern Atlantic provinces were sympathetic to the Nazi cause and may even have allowed U-boats to land and spies to disembark. Therefore military activity on both sides was rife in the region and it is not surprising to find that there is at least one spectacular sea-monster sighting from the area during the war years.

Again we are indebted to Dr Bernard Heuvelmans for the following little-known account :

'Not until 1962, when he published his Monsters of the Deep, did Thomas Helm tell about what he met in 1943 in St Andrew's Bay off the north-west coast of Florida in the Gulf of Mexico.

'After serving four years in the U.S. Marines, Helm was wounded at Pearl Harbour by seventeen bullets from an aircraft's machine-gun, which shot away his kneecap and part of his left hand. He was therefore invalided out, and at the time of our affair he and his wife were an-

nouncers at Panama City radio station. They were very keen sailors, and one day in March went out in a little 18 foot yacht along the west coast of Florida. The sea was almost as smooth as a mirror, when suddenly, soon after four in the afternoon, they saw a strange creature making straight for the boat. It had 'a head about the size of a basketball on a neck which reached nearly four feet out of the water.

'Helm steered a course to avoid this odd animal, but they still passed it quite close by:

'It was unmistakably some kind of animal. The entire head and neck were covered with wet fur which lay close to the body and glistened in the afternoon sunlight. When it was almost beside our boat the head turned and looked squarely at us. My first thought was that we were seeing some kind of giant otter or seal, but I was immediately impressed by the fact that this was not the face of an otter or seal.

'Helm had always been keen on zoology, had trapped otters and mink when he was young and had travelled over the north-west Pacific, so he had learnt to recognise sea-animals. The head of this creature, with the exception that there was no evidence of ears, was that of a monstrous cat. The face was fur covered and flat and the eyes were Set in the front of the head.

'The color of the wet fur was uniformly a rich chocolate brown. The well-defined eyes were round and about the size of a silver dollar and were glistening black. There was evidence of a flattened black nose and just below, where I judged the mouth should be, was a moustache of stiff black hairs with a downward curve on each side.

'For an instant the animal stared at the boat, then it turned away as if quite uninterested. Suddenly it swung round, dived and disappeared entirely. Only a big swirl of foam and an eddy of water showed where it had sounded near the boat.

'Helm, like other sailors, was interested in the problem of the sea-serpent, but was quite bewildered by the animal's appearance. A great dragonlike head with tooth-studded jaws would be much easier to explain away than a catlike head as large as that of a Bengal tiger. [But] what we saw did not in any way conform with reports I had read about unidentified sea creatures.

'He cannot have read many, for the animal reminds one of several other descriptions and seems to be a smaller, more round-headed-in other words a young-specimen of the maned merhorse, a sea-foal as it were.

'Helm's first thought was that it must be a sort of pinniped, but he rejected this at once, for he believed no known pinniped had so long a neck (which is true), nor so big a head (which is not, for the sea-elephant has); and that there were none in the Gulf of Mexico since the Caribbean Monk Seal, or West Indian Seal, was exterminated nearly 200 years ago (this is far from proved). There were certainly some at the beginning of this century, and there are good reasons to think that there are still some in the Bahamas, off Jamaica and the islands off Yucatan. But Helm also mentions a feature which seems to have struck many people who have seen maned sea-serpents:

'Seals and sea lions have long pointed noses and the eyes are set on the sides of the head like those of a squirrel or rat. The creature my wife and I saw had eyes which were positioned near the front of the face like those of a cat."

One of the most well documented series of sea-serpent sightings are those which have been recorded for centuries of the coast of British Columbia. Zoologists Ed Bousfield and Paul LeBlond have even taken the almost unprecedented step of giving the creature a scientific name *Cadborosaurus willsii* based upon two very blurry photographs taken in 1937 of a 12 foot long carcass taken from the stomach of a harpooned sperm whale. Bousfield and LeBlond have flown in the face of scientific orthodoxy by describing a zoologically improbable animal and claiming that it is a new type of marine reptile.

The truth is that, although there have been so many sightings that it is almost impossible to deny that *some* species of unknown animal (now popularly known as 'Caddy') is resident in these little known Canadian waters, it is hard not to agree with many less courageous scientists when they claim that Bousfield and LeBlond were precipitate in declaring that the creature is a reptile - let alone giving it a scientific name. Many people believe that the 1937 carcasses were the decomposed remains of a known species of animal.

However as Bousfield and LeBlond point out:

'Over the years, a number of other stranded carcasses have been equally disappointing. For example, a carcass found on Kitsilano Beach, in Vancouver, in 1941, dubbed "Sarah the Sea hag" by the press, was touted as possible remains of a Caddy. "She had a large horse-like head with flaring nostrils and eye sockets; a tapering, snake-like body 12 feet long; and traces of long coarse hair on the skin." The stinking remains were examined by Dr. W.A. Clemens, who had by then become a professor at the University of British Columbia, and his junior colleague, Dr. Ian McTaggart-Cowan. "We're not sure if it is a basking shark," said Dr. Clemens, "but there is no doubt it is of the shark family.'

However, G.V. Boorman, the former first-aid officer at the Naden Harbour whaling station (the scene of the discovery, three years earlier, of the mysterious carcass that Bousfield and LeBlond were to use as pivotal evidence in christening 'Caddy' as a new species of marine reptile), did not agree. Boorman was at that time a private in the army and had examined the stomach contents of at least 4,000 whales of various species over a period of ten years and claimed with some justification to be familiar with sharks in various degrees of decomposition or digestion.

'If that's a shark, I'll eat my uniform' said Boorman. *'I've seen the skeletons of scores of varieties of sharks, and they had no resemblance to these remains.'*

Another incident concerning what *might* have been the carcass of a hitherto unknown species of marine animal took place on the shores of Courock, on Scotland's River Clyde. Isummer 1942, an intriguing) carcass in an advanced state of rather smelly decomposition was found by council officer Charles Rankin.

Dr Karl Shuker, the British cryptozoologist, describes the carcass as....

"Measuring 27-28 if. long, it had a lengthy neck, a relatively small flattened head with sharp muzzle and prominent eyebrow ridges, large pointed teeth in each jaw, rather large laterally sited eyes, a long rectangular tail that seemed to have been vertical in life, and two pairs of "L"-shaped flippers (of which the front pair were the larger, and the back pair the broader)."

He notes that

"Curiously, its body did not appear to contain any bones other than its spinal column, but its smooth skin bore many 6-inch-long, bristle-like 'hairs' -resembling steel knitting needles in form and thickness but more flexible"

Rankin wanted to discover what this strange creature could be. He is quoted as saying that above all else . However, as this was the height of the war, and the locality was deemed to be a restricted area, Rankin was not allowed to take the photographs which would undoubtedly have allowed successive generations of zoologists to make a conclusive identification.

Unfortunately the remains were hacked into pieces, and buried. The only part which remained was one of what Rankin had described as *"knitting needle"* hairs that he kept in his desk as a *memento mori*, where it eventually shrivelled until it resembled *"a coiled spring"*.

As Karl Shuker writes:

"When considered collectively, features such as these bristles, the carcass lizardlike shape, vertical tail (characteristic of fishes), lack of body bones, and smooth skin suggest a decomposing shark as a plausible identity, but the large pointed teeth argue against the traditional basking shark explanation in favour of one of the large carnivorous species."

However, buried deep beneath what is now a Council Football Ground are the remains of a creature that could possibly solve the ancient mystery of the Great Sea Serpent.

As well as the putative Cadbarosaurus carcass Bousfield and LeBlond also note a number of sightings of 'Caddy' from the war years including one episode when the unfortunate beast was mistaken for a Japanese submarine (!):

'The war years were certainly a bad time for sea-serpents. Photographs or reports did not find their way into the press. But a few muffled echoes of Caddy were still to be heard. In April 1942 he was seen off Estevan Point. In the same year a Japanese submarine shelled this lonely point, and the superstitious fishermen maintained that they had been aiming at Caddy, having taken him for a secret weapon.

'Poor monster! Though the Japs might miss, a Canadian fisherman would not. Or so Ernest Lee, skipper of a motor fishing-boat, boasted when he claimed in the spring of 1943 that he had rammed the monster twice off Vancouver Island, and that Caddy had sunk to rise no more. This aroused much indignation in Victoria, and it was ironically suggested that Goering should spare one of his medals for the perpetrator of this wicked deed.

'The general silence about the sea-serpent during the grim war years was enough to make people think that it was well and truly dead. The disbelievers took this in the metaphorical sense that the world no longer had time for such nonsense, and the ignorant took it literally, assuming that the last specimen had finally been killed.'

During the war years, mysterious creatures were reported from the oceans all over the world; and interestingly there was a spate of Australian sightings at about the same time as the Japanese invasion of the island continent looked to be more and more imminent. Bernard Heuvelmans records that:

'.......in November 1941, two fishermen from Mooloolabak, north of Bisbane, saw from their boat a serpentine creature which they said was 60 feet long, it was immediately pronounced to be an oarfish. According to Charles Barrett, the naturalist who reported it, they said, 'Its body was marked with red and it had a red beard.' This does, indeed, agree with an oarfish, but these fish are never more than 21 feet long. Nobody suspected the fishermen of exaggeration, though this was certainly the case. And though the maned sea-serpent is often said to be reddish, nobody mentioned the monster's name. It was no longer done to talk about him.'

Within weeks of the attack on Pearl Harbour, American convoys sailed for the Pacific theatre of war. The 52nd Evacuation Hospital left with a large contingent from the hospital staff. Their destination was a small village on the island of New Caledonia, six hundred miles east of Australia.. They established a hospital in thatched roof and bamboo-framed huts totally constructed by hospital personnel with local civilian help. Hampered by obsolete equipment and lack of vital materials, the 52nd still treated over 6,000 inpatients and a larger number of out-patients during a six-month period. In the autumn of 1942, a forward unit was established in Normea. Within a few weeks, the small staff took care of 2,000 patients. The sea-plane tender, USS Curtiss and Patrol Squadron 14 arrived in the French colony, to begin operations from what became a principal Navy base in the South Pacific during the first year of the war.

Arthur Fere, was a trader at the time at Canala and twenty years after the war, as a direct result of the highly dubious 1965 sea serpent sighting claimed by French adventurer Robert le Serrec, he made the following report for Bernard Heuvelmans:

'As we were going into Ouengho Bay we all saw a strange shape, sticking up above the water. At first we thought it was a tree with a large branch pointing towards the sky, because it remained quite immobile ... The presence of this drifting object intrigued us very much, so we made for it. As we approached, we began to see a sort of big head followed by a black neck, marked with yellow. It reminded us of a giraffe.

'Following the neck, we could make out a big long shape just below the surface of the sea. We went on approaching. When we were about 200 yards away, it suddenly came to life and dived, raising a big plume of water. Our reaction was to put the helm about and make for land. I must admit we were all very frightened.'

Sea serpents are not the only mysterious denizens of the deep. The giant squid *Architeuthis* is one of the world's largest animals which is known to reach lengths of up to 60 feet (or about 18 meters); although animals much larger than that have been hypothesised. it is also one of the most mysterious and elusive creatures. There is more known about the prehistoric and extinct dinosaurs than these monsters of the deep. It has never been seen properly in its natural environment and the few sightings of living creatures have involved animals, which were dying.

Squid in the genus *Architeuthis* are found in the northern Atlantic, from Labrador to the Gulf of Mexico; northern Norway to the Azores; northern Pacific from the Bering Sea to the Sea of Japan: southern Japan, Hawaii and California. It is also found in the Southern Ocean. The best evidence indicates that the giant squids are cool water, deep sea creatures. They probably live at depths between 1000 and 3000 feet of depth. Like all squid, their blood is not highly efficient at carrying oxygen at higher temperatures. So they tend to live in open water where the prevalent currents are cool (ex. the North Atlantic current stream towards North America).

They eat other squid, fish and in the case of the largest, whales. This last fact is supported by eyewitness accounts. They are, in turn, eaten by sperm whales, though this is not as common as once thought. In fact, it appears that often it is the squid who will initiate attacks on the whale. This is borne out by the incident involving the Brunswick. The Brunswick was a 15,000 ton auxiliary tanker owned by the Royal Norwegian Navy - and in the 1930s it was attacked at least three times by a giant squid! In each case the attack was deliberate as the squid would pull along side of the ship, pace it, then suddenly turn, run into the ship and wrap its tentacles around the hull. The encounters were fatal for the squid; and since the animal was unable to get a good grip on the ship's steel surface, the animals slid off and fell into the ship's propellers.

Apparently, for some unknown reason, the Brunswick looked like a whale to the squids. This suggests that the sperm whale is not always the aggressor in the battles. At least three times, the squid swam beside and paced the ship before suddenly turning to attack. The squid could not grasp the metal of the ship and died when it slipped into the propellers. The giant squid is often seen as a likely candidate for the origin of the Norse tales of the Kraken, given that it inhabits waters sailed by the Vikings, and does on occasion attack ships.

One particularly chilling episode involving British sailors is recounted by Bernard Heuvelmans:

'Squids are predatory animals and will attack what they take to be their prey. This has, on occasion, included man. When the Britannia was sunk on 25 March 1941 in the middle of the tropical Atlantic twelve men were left adrift on a raft so small that some of them had to take turns at hanging on to it in the water. One night a huge squid hauled one of the men off the raft and he was not seen again. Soon afterwards a tentacle attacked Lieutenant R. E.G. Cox, fortunately letting go again, but removing discs of skin and flesh the size of a penny (one and a half inches diameter). This implies that the squid was 23 feet long. Two years later the scars were examined by Dr John L. Cloudsey-Thompson, the British biologist, and no doubt Cox

still bears them as proof of the ferocity of the squid.'

One wartime sighting even challenges the theory that the maximum size of this extraordinary animal is a mere sixty feet in length. One night during the War, a British Admiralty trawler was lying off the Maldive Islands in the Indian Ocean. One of the crew, A. G. Starkey, was up on deck, alone, fishing, when he saw something in the water:

'As I gazed, fascinated, a circle of green light glowed in my area of illumination. This green unwinking orb I suddenly realized was an eye. The surface of the water undulated with some strange disturbance. Gradually I realized that I was gazing at almost point-black range at a huge squid.'

Starkey walked the length the of the ship finding the tail at one end and the tentacles at the other. The ship was over one hundred and seventy five feet long!

Although practically nothing is known about its ecology, the Giant Squid is established zoological fact. The giant octopus, however, is a completely different matter. Known varieties of octopus range in size from a circumference of a few inches to as large as 23 feet. There is some evidence that, deep in the sea, there lives an unknown species of octopus that can grow to over a hundred feet across and weigh 10 tons! Only one colossal octopus carcass has ever been found and it was, and still is, surrounded by controversy.

However, a number of eyewitness reports suggest that a giant octopus is the creature behind reports of the lusca - an animal that has terrified the people of The Bahamas and the Gulf of Mexico for years. French cryptozoologist Michel Raynal has collected together a large number of giant octopus reports including this one from the late John C. Martin (US Navy, Retd):

'In 1941 I was a coxswain in the first division aboard the USS Chicopee AO-41... The ship had departed Baton Rouge, Louisiana, with a cargo of aviation gasoline and fuel oil for Portland, Maine. It was at the end of March or April that the ship was steaming of the coast of Florida in the general area of Fort Lauderdale and St. Augustine. Dead ahead of our course appeared something on the surface of the water that could not be readily described. The closer we approached it looked like a huge pile of brown kelp seaweed. As it hove into view there was no doubt as to its identity. The coils of its arms were looped up like huge coils of manila rope. However, the coils were over 36 inches [91 cm] in circumference.'

As Raynal has pointed out the measurements given for this creature - an estimated 30 feet [9.14 m] in diameter with arms that *'seemed about equal length; coiled but moving slowly'* is comparable to the descriptions of the animal washed up at St Petersburg, Florida in the 19th Century which to date at least, is the only specimen that we have of this enormous and wonderful creature. Raynal notes that: *'The "coils" of the arms were 36 inches (91 cm) in circumference, i.e., 29 cm in diameter, the same order of magnitude as the arms of the Florida monster.'*

Such creatures as the giant squid and the giant octopus are so ill-known by contemporary science that they have achieved the status of monsters in many people's eyes. However there is no doubt that they exist.

More problematical, at least to mainstream science, are the accounts of huge creatures from inland lakes and rivers across the globe. Again, the Second World War spawned a number of sightings of such beasts. Probably the most interesting is of a thirty foot long, humped creature seen in the mid 1940s in Lake Ikeda on the Japanese island of Kyushu, and a 1942 sighting by two Soviet pilots of two creatures described as being like 'giant newts' in Lake Yautia in Siberia.

As we have now demonstrated, there is clear and persuasive evidence to show that encounters between the world's military powers during the War and mysterious beasts proliferated. However, a remarkable - and ingeniously original - theory has been put forward to explain the repeated sightings of a strange creature in an American lake in 1944, and it is a theory that has a direct hearing on distinctly covert wartime activities of the US Navy!

According to US college teacher Jim McCleod, in the war years reports began to surface from Lake Pend Oreille of an elusive beast dubbed "Paddler". However, when McCleod began investigating the eye-witness accounts, he was intrigued to note that sightings of the monster coincided very nicely with secret US Navy submarine tests that were being undertaken In the lake!

The 'monster story', McCleod feels, served

(a) to deter people from wanting to visit the lake (maybe, but it wouldn't have deterred us - quite the opposite, in fact!);
and
(b) to ensure that anyone who might have seen unusual underwater activity at Lake Pend Oreille would have been labelled a crank!

As McCleod puts it: *'I think the Navy has directly or indirectly perpetuated the monster rumour for some time to disguise what they had been doing in the lake. They did not want people sticking their noses in.'*

Despite the fact that the Navy denies the existence of any such *operation ('I think that it is definitely false,'* stated Commander Rick Shultz, head of the Navy's David Taylor Acoustic Research Centre, which is situated at the southern tip of the lake), in all probability McCleod is correct: in the 1970s, sightings of Paddler once again surfaced - at precisely the same time, it was revealed a decade later, that the Navy was testing out a new design of torpedo in the lake...

The best-known lake monster in the world, however, is 'Nessie', the denizen of the Great Glen in Scotland. Loch Ness is located in the north of Scotland and is one of a series of interlinked lochs which run along the Great Glen. The Great Glen is a distinctive incision which runs across the country and represents a large geological fault zone. The interlinking was completed in the 19th century following the completion of the Caledonian Canal.

Many lakes of Northern Scotland had ancient legends about monsters and the like. In 565 A. D., though, Loch Ness's story was written down. The account tells of Saint Columba who saved a swimmer from the a hungry lake monster. From then on, rumours about the creature were repeated from time to time; and in 1933, after a new road was build along the edge of the Loch, the number of reports soared.

There were a number of reports from the War years, the first of which comes from the collection of Nicholas Witchell, who, perhaps better known as a BBC newscaster, has made a study of the mysterious denizens of Loch Ness:

'During the Second World War members of the R.O.C. were stationed around Loch Ness. It is said that several members saw the animals but were forbidden to report the fact because of military discipline.

'One member, however, did make his sighting known to a close friend. At 5.15 a.m. on 25 May 1943, Mr C.B. Farrel, while on duty at Fort Augustus, saw an unidentified object on the loch. When he looked through his binoculars (Zeiss x 6) he saw a creature twenty-five to thirty feet long about 250 yards away. In colour it appeared to be dark olive brown on top and lighter underneath. Its eyes seemed to be large, and the neck was described as being graceful and four to five feet in length. It was evidently feeding, since it kept depressing its head and neck until they were submerged and then it would quickly withdraw them from the water and shake its head vigorously. In the end the whole body slid out of sight without causing any disturbance.'

Another account from Witchell's collection would seem to have come from an unimpeachable source.

'The Deputy Lord Lieutenant of Inverness-shire is Mr William Mackay, D.L., O.B.E., F.S.A He is also the Chieftain of the Glen Urquhart Games and used to be Dean of the Faculty of Lawyers in Inverness. Throughout his life he has led an active outdoor existence, which has included two sightings of the animals. At his home in the beautiful Strath Glass, near Loch Ness, he gave me this account. The (second) time Mr Mackay saw it, he told me, was just before the end of the last war. He was driving home from Foyers one evening when, about six hundred yards away across the loch opposite Urquhart Bay, he saw the same two humps again. He went on:

'Fortunately this time I had my deerstalking telescope with me and so I stopped and examined the 'monster' through my glass. It appeared to be about thirty feet long in all, with a long neck which it kept flat on the water. There were two humps which were dark elephant grey ill colour. It looked as though there was hair over its back and body. The wind was from the west and the beast seemed to be trying to keep its head on to the wind, because every now and then I saw splashes and a long tail appeared to be sculling and two flippers to be paddling to change its position. Before I left, a Mr Deans, a plumber, drove up and told me that he too had been watching the animal.'

Although many cryptozoologists believe that the creature of Loch Ness is a flesh and blood animal - either of a species new to science or, more probably, of a known species which has grown to an unprecedented size, other fortean researchers following the lead of such pioneers as F.W. Holiday believe that the phenomena have a more paranormal explanation. There have been numerous ghost stories, UFO reports and accounts of ritual magick in the area, and it seems almost certain that some, at least, of the sightings of 'Nessie' have a more arcane explanation than the mainstream cryptozoologists would like to accept.

Lending weight to this argument is a story from author Bruce Barrymore Halpenny:

'I was at the Loch early one morning, at the crack of dawn, camera at the ready," said Pete Smithson, "suddenly I saw this figure coming, weaving like, towards me. I thought he had been in some sort of accident... I then saw he was dressed in wartime flying clothes, complete with parachute and harness.

'As he sort of staggered towards me I suddenly felt cold...By now he was no more than 10 to 15 feet away and I could clearly see his RAF uniform. I shouted, 'Are you alright?' At that point he arched his right-hand and pointed towards the Loch.

'I Instinctively turned and looked out over Loch Ness in the direction In which he was pointing...l fully expected to see the Monster. I then had a funny feeling. The coldness had gone and, quickly turning to the strange? shouted - 'What is it?'. But he had vanished. I then realised that it had been a ghost airman. But, he looked so real.. the only thing about him was he looked injured and, his face was greyish.

'What a damn fool I felt, confronted by a ghost, my camera around my neck, yet I never had an inkling to take a photo.'

'That sighting was in September 1978 and since then, two other people have also Informed me that they had seen a figure of an airman near Loch Ness which, had vanished when near the Loch. They both said the figure had on parachute and harness.'

Barrymore goes on to explain that :

'...a Vickers twin-engined Wellington, N2980 A-Robert, crashed into Loch Ness on New Year's Eve, 1940. Caught in a snow storm one engine cut out and, unable to maintain height, the pilot, Squadron Leader Marwood-Elton, ordered the crew to bale out. They did so but, the rear-gunner's parachute failed to open properly and he was killed. The Wellington bomber crashed In Loch Ness, the pilot and co-pilot having remained at the controls to keep the bomber steady while the crew made good their escape. They were unaware that the rear-gunner had been killed. Before the bomber sank into the Loch, they managed to scramble into the dinghy and reach the safety of the shore.'

Barrymore hypothesises that the air-gunner was trying to give some sort of warning to the pilot and notes that in his opinion at least it is highly strange that the sightings of the Phantom World War Two flyer, only started in 1978:

'...a year after Robin Holmes of Heriot-Watt University, Edinburgh located the wartime bomber. For 37 years Wellington N2980, 'A-Robert', had remained undisturbed. Now, the bounty hunters were in full cry, for out of the 11,461 Welllngtons built, this was the only survivor. Yes, there had been hundreds, Germany had not destroyed them. But, after the war they were not wanted and melted down ... no-one was interested. Now everybody was interested.

'In September 1988 they started to lift A-Robert and all was going well...when suddenly, something gave...the bomber sank back into the Loch. Was that a sign to call a stop? Leave the Wellington bomber in peace...was that what the phantom flyer was trying to say?

'But they did not stop and after another attempt the Wellington bomber was brought out of the Loch ... the tyres were still inflated and when a battery was put on, the navigation lights worked at once.'

Today, Wellington N2980 is at the Brooklands Museum being restored, and although Barrymore asks whether *'the phantom World War Two flyer is at peace now his bomber had been removed from its last resting place'*, the greater question remains. Despite a widespread human belief in ghosts and phantoms, there is minimal scientific evidence for their existence, and the laws of physics, which much define their reality, remain unknown. Out of all the thousands of dead bomber crews from the Second World War why is it THIS one that remains on earth to haunt future generations? Is it because of the peculiar tragedy of the young men's death or could it perhaps be because Loch Ness as a geographical area is one, which is particularly conducive to paranormal and psychic phenomena...like phantom airmen, UFOs and the Loch Ness Monster?

Terrors of the Taiga:
THE MONSTERS OF SIBERIA

PART ONE

by Richard Freeman

The world's mightiest forest is not the Amazon rainforest. Neither is it the jungles of the Congo. It is found far north of the tropics, a cold, verdant wilderness consisting mainly of pine and Siberian larch (the most numerous tree on the planet). The Russian Taiga is a forest of almost unbelievable dimensions stretching from the borders of northern Europe in the west across the north of mother Russia to the Bering Sea in the east. Shot through with freezing swamps it is almost entirely uninhabited. The Taiga covers an astounding 7 million square kilometres.

The Taiga is at its wildest and most ill-explored in the icy abandoned region known as Siberia. Here winter temperatures drop so low that they can shatter steel. The brief summers are haunted by clouds of blood hungry mosquitoes that will cover any warm blooded animal. Siberia consists not only of the Taiga but endless miles of swampy tundra and mountains and plateau where no man has ever set foot. In comparison, the Amazon seems about as wild and untamed as a flower bed on a roundabout in suburban Dorset.

It comes as no surprise that such an unknown land has produced reports of monsters. Such a vast area of the planet unspoilt is bound to be home to unknown species. But the stories that

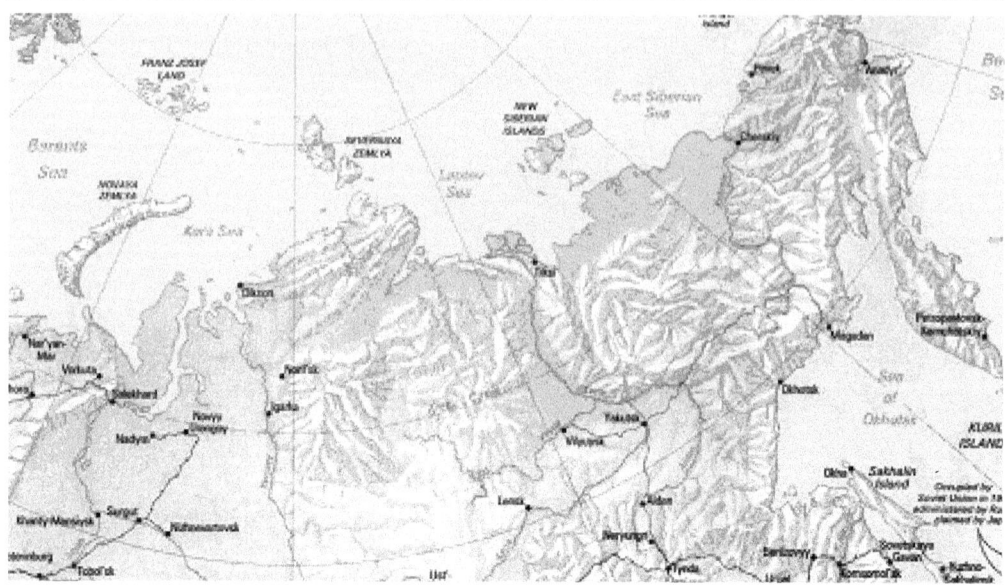

emanate from the Taiga and its environs seem altogether stranger than their analogues elsewhere.

THE SIBERIAN SNOWMAN

It is fitting that this lost world has its own ape man. It has many names in Siberia, *albasty, kiik-kish, chuchuna, kuchena, kul, mulena*.. Such creatures, resembling hulking bipedal apes, are reported from dozens of countries in every continent, but it is only in Russia that the scientific establishment has had the common sense to take these creatures seriously. The late Pytro Smolin, curator of the Darwin Museum in Moscow, began organizing the Relic Hominoid Research Seminar until his death in 1975. It is hard to imagine a western scientist in Smoiln's position doing this. The small minded, arrogant, whining of armchair zoologists would be deafening. In Russia hominoid research has a long and venerable history.

Dr Boris Porshnev one of the world's foremost authorities on Russian hominoids has collected reports from all over the former Soviet Union. Siberia, the most untouched region, is rich in snowman tales. A.P Okladnikov, an archaeologist who worked along the lower Lena River, told Porshnev of his findings.

"The Chuchuna are a tribe if half-men, half-animal beings, still occasionally met with in the North. The creatures have no neck and heads that consequently seem to sprout straight up from their torsos. They usually appear at night, unexpectedly, and throw rocks on the sleeping humans from the cliffs. They are given to trapping reindeer. A Yakut hunter named Markarov said he found caves inhabited by the creature on the River Lena's right bank and as far as Lake Stolb. In these lairs were many antlers, and some hides of the reindeer that had been eaten."

Back in 1912 P. L. Darvert, a young mineralogist, published reports of hairy, wild men he had witnessed since 1908 along the lower Lena. Later he became a Professor specialising in the study of Meteorites but then reverted to his earlier studies. In 1933 he wrote a long paper on hominoids *The Mulen and Chuchuna Wild Men.* One Yakut tribesman told Darvert that the chuchuna sometimes crossed from the Lena to the Aleutian islands. This is interesting as all known apes are poor swimmers and sometimes drown when they fall into deep water in captivity. The tribesman recounted how one of the monsters was found lying on the seashore one day. No one dared approach, as they did not know if it was dead or not.

Even the dogs would not go near it. The beast lay unmoving all day. Only one tribesman kept watch over it after sun set. He saw it rise from the ground and make off..

Porshnev's death in 1972 prevented him from formally writing up his "polar chapter" in snowman research, but others have followed in his footsteps. One of the most dedicated was Vladimir Pushkarev, a bold young biologist, who often braved the wilds of Siberia alone. On these solo expeditions he hoped to get closer to the Siberian yeti than with large expeditions consisting of many people. In an area such as Siberia this took amazing bravery. Pushkarev was working on a thesis entitled *Current knowledge on the Relict Hominoid in the North of Eurasia.* Tragically Pushkarev was drowned whilst on a one-man expedition to the Ob River basin and the surrounding swamps and forests in 1978, the ultimate sacrifice in the name of cryptozoology. His body was never recovered. What follows are some of his findings from the area that was later to claim his life. Luka Tynzyanov, a former Taiga hunter, told him:

"In 1960, or it might have been 1961, I was hiking one evening from yarskogort to Vasyakovo along the bank of the Gornaya Ob. I had two dogs with me. Suddenly they bristled, began to bark and ran ahead. They came running back and this time huddled near my feet and stopped barking. Just then two kuls emerged from the forest. One was tall, over two meters, and the other a little shorter. I also got scared because their eyes glowed like two dark red lanterns. They came towards me and, when they were quite close, stared at me with flashing eyes. They wore no cloths. They were covered in thick short hair. Both their faces and bodies were black. Their faces jutted out, their arms were longer than a man's and they swung them in a strange way. Their gait was unlike a human being's. They turned their feet in when walking. When they passed us, the dogs made a bee-line for the village.

Tynzyanov had seen two other specimens shortly after WW2.

In Salekhard a former school teacher called Marfa Sekina, who taught tribes people to read and write, told Porshnev of her encounter with a snowman in her youth.

"Before the Revolution my father and I were constantly travelling around the northern Ob region and the Yamal Peninsula. I was 20 at the time, and our permanent residence was in Salekard. Sometimes we stayed with an old Khanty not far from the village of Puyko.

I remember that it was in September, the nights were dark and our dogs bayed at night. Once their barking was particularly ferocious. The next night it was no less frenzied. I asked our host, the Khanty, whom they were barking at like that, and he whispered that it was the Zemlemer (land surveyor).

Zemlemer? I was puzzled.

I'll show you tonight, he said. Only watch him with caution - through your fingers. At midnight we walked out of the choom (a tent of skins and bark). There was a large red moon. We waited for about an hour, and suddenly the dogs began to bark. Several dozen meters away I spotted a very tall man. Our chooms were surrounded by a hedge of rose willows two meters high.. The man's head and shoulders rose above it. He walked fast, with long steps, pushing right through the thickets. His eyes glowed like lanterns. I had never seen such a tall and terrible man. The dogs rushed at him, baying. One, lent courage by our presence, ran right up to him. The man bent over, picked it up and hurled it far to one side. We heard a yelp and saw the dog's body careen through the air. The man left quickly and did not turn back to look at us..

Was it a forest sprite? I asked the old man.

Don't say that word he said in fright, lest you summon him. Just call him Zemlemer. He comes here, every year at this time.

Next morning one of our dogs was missing."

The glowing eyes and antipathy towards man's best friend are features of hominoid reports from all over the world. Pushkarev questioned pupils at three specialized secondary schools in Salekhard. The children were all from the Yamalo-Nenets National Area and belonged to reindeer breeding families. 48 out of the 60 questioned had come across the wildman in the tundra. All 60 knew that the Nanets called it Tungu. Four of them had seen it recently (in the sixties or seventies) but at a distance and in twilight. A further 10 said that their relatives had seen it. Those who had seen the *tungu* described it as tall and shaggy with a fast run and a shrill whistle.

In the village of Nyda in the Nyadm district he questioned reindeer breeders who all believed in the *tungu* but who said it had not been seen for 10 years. In the early sixties it was seen quite often.

In 1974 Pushkarev conducted research some 5,000 kilometres away from the Ob River in Yakutia in eastern Siberia. A story recounted by Tatyana Zakharova, a 55 year old Evenk man, took place on the bank of the Khoboyotu Creek.

"After the revolution, in the twenties, our villagers came across a Chuchunaa while out berry picking. He was also picking berries and stuffing them into his mouth with both hands. On catching sight of us, he stood up to his full height. He was very tall and lean, they say. Barefoot and dressed in deer skin, he had very long arms and a mop of unkempt hair. His face was as big as a human's . His forehead was small and protruded over his eyes like the peak of a cap. His chin was large and broad, far bigger than a human. The next moment he ran away.

He ran very fast, leaping high after every third step."

This last account seems to describe something different to the others. This being seems more human and less ape like than the other reports. It seems to a feral human, perhaps part of a very primitive hunter–gatherer tribe, not a true snowman at all. It lacks the body hair, muscular bulk, and glowing eyes of the genuine article, These ultra primitive humans or pseudo-snowmen are written about in Soviet historian and ethnographer G. V. Ksenofontov's book *Urankhai Sakhalar.*

"The Chucunaa is a human. He feeds on wild deer and eats the raw meat. They say he tears the skin off the wild deer and wears it, just as we do the hide of a fox. He lives in a lair like a bear. His voice is unpleasant, grating and hoarse. He whistles, frightening people and reindeer. Men come across him very rarely, and often see him running away. The Chuchunaa's face is black. It is hard to make out the nose and the eyes. He can be seen only in summertime. In winter he is not around."

Confusingly, two separate entities, one human, one ape or ape-man, are referred to as a *Chuchunaa*. The next reports clearly refer to something *not* human.

Another tireless hominoid researcher is Maya Bykova. Her perseverance has rewarded her with some of the closest sightings of the Russian yeti to date. In 1985 whilst on her way back from a trip to western Siberia she met a young man on her train and fell into conversation with him.

The fellow, Volodya, was an ethnic Mansi of western Siberia and found Bykova's studies very interesting. Once he had established that she did not want to shoot hominoids he confessed that he, his grandfather and his father, had seen such a creature in a hunting lodge. The lodge was located in a cedar forest surrounded by bogs some 70 kilometres from his home.

The creature was large, bulky and covered in dark reddish-brown hair except on his left forearm, where the hair was white. For this reason they named him Mecheny (Marked). He would always knock on the cabin window and then wander around the building as if looking for something on the ground and muttering. Their hunting dog always ran away when he appeared and returned when he left. Mecheny was only seen in August and had been spotted twice in 1985.

Volodya kept in touch and wrote to tell Bykova that Mecheny had returned three times in 1986. He extended an invitation to Bykova to join his family at the lodge in 1987. In August of that year she travelled to the lodge with Volodya, his wife Nadya and a three month old puppy called Box (the old dog having passed away). At dawn on the first day they were awakened by two sharp knocks on some plywood outside the cabin. Bykova rushed out to see what had made the noise. Her hosts followed, worried for her safety.

"It was dawning and the first thing I saw in front of me was a white spot against the dark background of trees. After that I saw his figure. He was standing five meters away, his right shoulder leaning against the barkless trunk of a dead cedar. Sharply in view were the white

forearm and brightly glowing red eyes. It was sufficiently light and I was close enough to see him in detail. He stood two meters, give or take five centimeters. Looking down at us (Voloydya is 180 centimeters tall, I am 168, Nadya is shorter), he shifted his glance from one to another and made a sound, something like "Khe!", as if clearing his throat without parting the lips. On the whole, judging by his build, especially his lower extremities, he resembled a man, not an ape or a bear standing on the hind legs. But like an animal he was covered all over in fur, some six or seven centimeters long, red-brown in colour, except the left forearm, as already mentioned, was white.

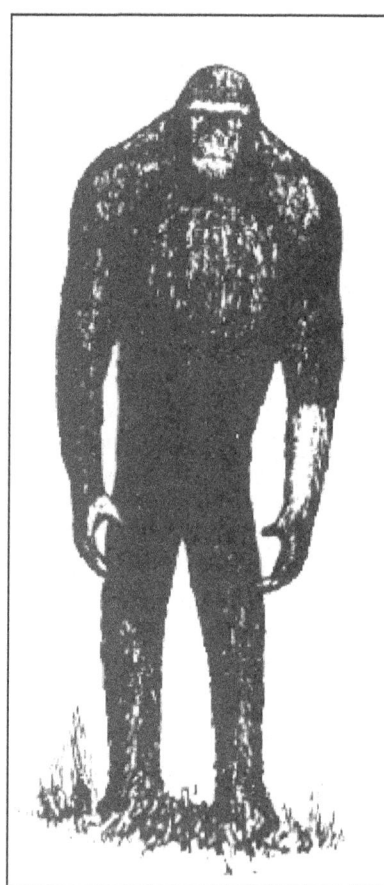

A drawing of Mecheny from the description by Maya Bykova

I drew the creature's portrait from head to foot as it stuck in my memory. The head, facing me, looked round, but later, when he turned, I noticed that the back of the head was elongated. The hair on the head was short, no more that three centimeters. I did not see any skin on the face, it was all covered with hair, including the ears, the nose, and the nostrils. I could only see the eyes, almond-shaped and like a man's but sunken under prominent brow ridges.. The jaws were slightly put forward and showed a long narrow slit of a mouth. The head sat right on the shoulders, without a neck. The shoulders were strikingly wide and strongly muscled. Such musculature in humans can only be seen in body-builders. The chest was powerful and barrel-like. Hefty arms, set somewhat forward, hung loosely down. Their relative length seemed within human proportions. The hands were enormous and shaped like scoops. I could see the skin of the palms and it was reddish. In the groin the hair was longer, the genitals were not seen. The legs were long and straight, with enormous feet. They were also covered with hair and I did not see any skin."

Bykova counted the seconds off in her head as the trio stood watching the snowman. At 60 seconds Box came out of the cabin and snarled at Mecheny. The monster turned and vanished into the forest. Over the next eight days he did not return and Volodya had to return home as his leave was running out.

It is interesting to note the coughing sound Mecheny made. Gorillas will make such a sound as a warning or threat. Was the snowman doing the same thing? Events would soon prove that Mecheny was far from harmless. In mid October, Bykova returned to the cabin with Volodya, Nadya and Box (now five months old). This time nothing knocked on the window or wall but one night they heard a strange cry. The following morning Box was missing and they set out to find him. The came upon his body some 100 meters from the cabin. Poor Box had been torn apart from the tail to the clavicals. The right side of his skull had been crushed with such force that his teeth had pierced his tongue. It seemed the snowman had stalked Box, seized him by the legs and smashed him against a tree with superhuman strength. Volodya, who was a seasoned bear hunter and woodsman, became worried and the team packed up and travelled home. This would not be Bykova's last sighting of Mecheny. It may seem odd that Byakov as a scientist did not take a camera. She was afraid that it might scare the creature. Her whole approach was to habituate Mecheny until he was used to human contact then film and photograph him.

Encounter with Mecheny based on witness description

In August of 1988 she returned to the forest with Volodya and his grandfather. They spent their nights concealed in the trees on the edge of the clearing in which the cabin stands. After they had no luck they moved some 800 meters from the cabin to the edge of the swamp. On the night of the 22nd a summer storm brewed illuminating the sky with lightning flashes. At a distance of seventy meters a flash lit up a hunched figure. It rose revealing a white forearm. The creature seemed to be jumping around and falling to the ground. As he rose he brought his right hand to his mouth and seemed to be eating. The watching humans followed his movements as he came within 25 meters of them. He was apparently hunting small animals, perhaps frogs. His nocturnal hunt was the only time he was observed on this trip.

Russian hominology seems to be constantly tinged with tragedy. Byakov's plan to habituate Mecheny never came into fruition. Volodya's wife Nadya died suddenly at a young age and his grandfather passed away a few months later. Devastated, Volodya wanted only to be left alone. Letters of condolence were answered with thanks, but the young man was, understandably not interested in further research. He had always been worried about "giving away" Mecheny and having the forest over-run with hunters. Thus, one of the most promising episodes in cryptozoological history was brought to a melancholy end.

Moving to eastern Siberia once more Reuters ran the following report August 16th 1990.

"Soviet KGB border guards were put on a state of maximum alert after a night patrol sighted a creature 2 meters tall with brilliant eyes. The beast, which surprised the guards of the Red Flag border post in the Soviet Far East, resembled the mythical Abominable Snowman, or Yeti. Soon afterwards it was seen trying to climb up to a roof, but eventually retreated to the forest."

The use of the word mythical in this report is clearly in error. Dmitri Bayanov successor to Boris Porshnev was told of an earlier encounter by a border guard Constantine Shemberev that took place in the winter of 1983.

"This happened not far from the city of Birobidzhan in a forest on the Chinese border. My driver, an ethnic Uzbek, and I were spending a night outdoors. It was very cold and we made a fire. I left him looking after the fire and went to collect firewood. As there was much snow, it took me about half an hour to collect an armful of brush. Suddenly there was a blood-curdling cry from the fire-pit. A cry like that on the border is especially alarming. I dropped the firewood, took my tommy-gun at the ready, and ran to the camp site. Suddenly I stopped, seeing a beast heading in my direction.

It was very big and I was taken aback. As for the beast it showed little fear. When there was only about three meters remaining between us, it straightened up and I saw it was a man. He was very shaggy and in some places the hair was matted into plaits. The eyes were human-like and he stared at me. I thought that if he made another step I would shoot. But he made a sound like a chuckle and went aside. Walking away, he turned around a couple of times as if to make sure I was not following.

He was stooping very much but his arms were not hanging down like those of an ape. Then I remembered my driver and ran towards the vehicle. The driver was hiding behind the bonnet and was very glad to see me. He was trembling excessively and it took me about an hour to calm him down. He told me the following story: He was tending to the fire when that thing stepped out of the forest. The driver became petrified with fear and could not make a step. The shaggy man approached the fire and started throwing snow onto it. The driver noticed his very long fingers. Suddenly the "man" heard the crunching of snow under my boots and turned in my direction. At that moment, the driver let out a piercing cry and dashed towards the automobile, while the "man" headed in my direction. In brief, that's the whole story. Constantine added further details in letters to Bayanov. The beast was two meters tall with wide shoulders and dark coloured hair. The forehead stuck out, the nose was wide and flat, the neck was not visible. The driver remembers the long fingers and big eyes. The snowman had come within two paces of him. As a Muslim he believed it to be a devil. Constantine thought it was an ape or wildman.

In 1988 Bayanov was sent a copy of the magazine *Magadan-skaya Pravda* from a colleague living in the northern Russian city of Magadan on the coast of the sea of Okhotsk. S. Kozlovsky, author of an article in the magazine , writes that in 1979, as part of an educational programme he lectured to reindeer herders on the origin and evolution of man. His lectures were accompanied by slides and given via an interpreter. When a slide showing a reconstruction of *Homo erectus* was shown the audience all identified it as a *pikelian*. This was repeated with every new group he lectured to.

The said the *pikelian* was a man-like creature covered in grey brown hair. One hunter named Mikundyya was said to have come across a female specimen in the mountains.

He watched from behind a boulder as she dug up a root, cleaned it and ate it. He decided to capture her and grabbed her from behind.

She let out a scream and dragged him along until he banged into a rock and let go. The same man was also supposed to have come across a *pikelian* cave with a bed of grass and moss, and lots of small animal bones. Further east still on the Chukchi peninsula Russian Victor Chebortarev saw a snowman in 1970.

During a hunt on the Amguema River, he and two companions sighted a gigantic hairy figure that seemed both man like and ape like. It bore wide shoulders, a small head and hefty arms and legs. It stood motionless the turned and disappeared behind a rock.

What is the Siberian snowman? Ape, man, or neither? The giant hominoids of China and the Himalayas, the *yeren* and *yeti,* may be a surviving form of the giant Pleistocene ape *Gigantopithecus blacki*. This ape is known from fossil teeth and jaws dating back 500,000 years.

It was probably a biped and would have stood 3 metres or more tall. It fed on bamboo and fruit. Fossil seeds of the durian fruit have been found in its teeth. But *Gigantopithecus* seems to have been a creature of the tropics and sub-tropics. The Siberian snowman inhabits some very cold areas. It also seems smaller than the tropical giants. Dmitri Bayanov thinks it may be an offshoot of Neanderthal man. He postulates that these beings were far larger than true Neanderthals, had smaller brains and were more hairy.

They probably lack fire and create only rudimentary tools. We know true Neanderthals could cope with extreme cold. Their more bestial relatives are likely to be even more hardy. It is interesting to note that these creatures have been reported from European Russia, close to the border with Finland. Could they have been behind the genesis of the Scandinavian troll legends? I will leave the final word on the snowman to the bold Vladimir Pushkarev.

"The problem of relict hominoids is one of the great enigmas of earth. Its importance is at its peak today because in a decade or two these relicts may disappear from the face of the planet, just as did the mammoth at the time of the Roman empire and Epiornis [sic] (the gigantic bird of Madagascar) in the 18th century. We are the generation that finds the hominoid still alive and that is why we are fully responsible for the solution of the problem.

SURVIVORS FROM THE ICE AGE

The archetypal beast of the freezing north is the mammoth. Next to the dinosaurs this must be the most familiar of all pre-historic beasts. We have many misconceptions about the mammoth. We tend to think of it as one species, the woolly mammoth (*Mammuthus primigenius*), when in fact there were many. These ranged from the 20-ton imperial mammoth (*Mammuthus imperator*), to the tiny race of dwarf woolly mammoths on Wrangle Island in the Siberian Arctic Ocean. We also associate the mammoth with the bleak open tundra but this large herbivore would need vast amounts of vegetation to support it. The taiga was its most likely home. In Siberia the woolly mammoth became extinct around 12,980 years ago. On Wrangle Island they lingered to 4010 years ago.

Tribesmen of Siberia have long known of the mammoth, calling it 'the animal that sleeps' Many believed it was a burrowing animal that died if it came to the surface. This was due to the numerous specimens preserved in the permafrost. Occasionally forced to the surface, these would seem like cthonian animals to the uneducated. So well preserved are some mammoth that their flesh is still edible. One famous specimen unearthed by an expedition in the early 20th century was so fresh that the sled dogs ate its buttocks! Hence the museum mounting has it in the odd position of sitting down.

We do not know if it was climatic change, hunting by humans, or a combination of both that drove the mammoth to extinction. But the intriguing question remains, do any still roam the world's largest forest unseen by man?

A Russian hunter claimed to have seen a pair of mammoths in 1918. He told his story to M. L. Gallon who was in charge of the French Consulate at Vladivostok.

'*In the second year I was exploring the taiga, I was very much struck to notice the tracks of a huge animal, I say huge tracks, for they were a long way larger than any of those I had often seen of animals I knew well. It was autumn. There had been a few big snowstorms, followed by heavy rain. It wasn't freezing yet, the snow had melted, and there were thick layers of mud in*

the clearings. It was in one of these big clearings, partly taken up by a lake, that I was staggered to see huge footprints pressed deep into the mud. It must have been 70 cm across the widest part and 50 cm the other way, so the spoor wasn't round but oval.. There were four tracks, the tracks of four feet, the first two about 4 metres from the second pair, which were a little bigger in size. Then the tracks suddenly turned east and went into the forest of middling-sized elms. Where it went I saw a huge heap of dung; I had a good look at it and saw it was made up of vegetable matter.

Some 10 feet up, just where the animal had gone into the forest, I saw a sort of row of broken branches, made, I don't doubt, by the monster's enormous head as it forced its way into the place it had decided to go, regardless of what was in its path.

I followed the track for days and days. Sometimes I could see were the animal had stopped at some grassy clearing and then gone on forever eastwards.

Then, one day I saw another track, almost exactly the same. It came from the north and crossed the first one. It looked to me as if they had trampled about all over the place for sev-

eral hundred metres as if they had been excited or upset by their meeting. Then the two animals set out marching eastward one following some 20 metres behind the other, both tracks mingling and ploughing up the earth together.

I followed them for days and days thinking that perhaps I should never see them, and also a bit afraid, for indeed I didn't feel I was big enough to face such beasts alone.

One afternoon it was clear enough from the tracks that the animals weren't far off. The wind was in my face, which was good for approaching them without them knowing I was there. All of a sudden I saw one of the animals quite clearly, and now I must admit I really was afraid. It had stopped among some young saplings. It was a huge elephant with big white tusks, very curved; it was a dark chestnut colour as far as I could see.. It had fairly long hair on the hindquarters but it seemed much shorter on the front. I must say I had no idea that there were such big elephants. It had huge legs and moved very slowly. I've only seen elephants in pictures, but I must say that even from this distance (we were 3000 metres apart) I could never have believed any beast could be so big. The second beast was around, I saw it only a few times among the trees: it seemed to be the same size.'

The hunter did not know what mammoths were. When Gallon spoke of them it was clear that the hunter was unfamiliar with the term.

An even more amazing story came to light in 1873 when the *New York World* published an interview between one of its correspondents and a Russian convict called Cheriton Batchmatchnik who had been pardoned by his government due to his remarkable discoveries. Batchmatchnik had been convicted of smuggling and had been banished to Siberia. He had been set to work in the mines of Nartchinsk but soon escaped. Having reached the mountains he struck southwards heading for the Amur river in the hope of reaching China. After meeting Cossacks he turned north and entered what seemed to be a pass in the Altai range. He got out to the north by a branch in the Lena river. Turning eastward he entered the gorges of the Aldan mountains. Here winter overtook him.

He followed vast herds of migrating animals to the summit of the range. There below him was a hidden valley, hemmed in by cliffs on all sides. He estimated it to be fifty miles wide by one hundred and fifty miles long. As he descended he found the valley to be warm and fertile. The place was filled with animals and contained a blue lake. He made camp and lit a fire beside the lake.

When nightfall came some huge animals approached, apparently attracted by the fire. They crushed willow trees like pipe-stems and made guttural bellows. Batchmatchnik fired his pistol into the dark, causing a stampede. Come daylight he discovered their massive spoor and a well-worn track leading to the water. He decided to seek a safer spot for his next night. He discovered a cave and on entering heard a deep breathing. There standing in the cave was a full grown mammoth. He described the beast as 12 feet tall and 18 feet long. It was covered in reddish wool and black hair. The curving tusks were 8 to 10 feet long.

In the coming days Batchmatchnik saw about twenty mammoths in the valley. All were adults and he saw no calves. They were peaceable animals who were never aggressive to him and indeed took little notice of him. There was however a much more dangerous animal in the valley. He saw it only once. It was a dragon-like monster that lived in the lake and preyed on animals that came there to drink. It was 30 feet long, and armed with savage fangs. It was covered with scales. It seems to have been the same species that we shall examine below which has been reported from several Siberian lakes. One morning he observed this "saurophidian" (as he called it) attack a mammoth. The reptile seized its victim and enfold it in its coils. After an hour long struggle the mammoth managed to pull itself free and limp to safety.

Batchmatchnik eventually left the valley and found his way back to civilisation. What are we to make of this story worthy of Conan Doyle or Edgar Rice Burroughs? It may seem fantastic but the Russian officials seemed to believe it as they pardoned Batchmatchnik in recognition of his 'services to science'. To my knowledge there were no follow up expeditions (as always seems the case). If this strange story is based in any way on the truth then a tumultuous discovery awaits us in the Aldan mountains.

The Ostiaks and Yakuts claim to have seem mammoths but I have no details of their reports. Yermak Timofeyevitch was the leader of a band of Don Cossacks employed to rout bandits plaguing the salt mines in Siberia in 1580. They ended up conquering the whole of Siberia. At the beginning of the campaign Timofeyevitch reported that beyond the Ural mountains he had met with a 'large hairy elephant'. The natives called such beasts the 'mountain of meat.' Was this just a defrosting dead mammoth or a live one?

No recent sightings have filtered through to the west. But the taiga is truly vast and may still hide mammoths. Even if the mammoth *is* extinct it may return from the dead to roam Siberia again. Professor Akira Iritani of Kinki University has perfected a way of fertilising egg cells with dead sperm. He has successfully tested this on farm animals. Iritani has come up with a plan worthy of a science fiction novelist. His audacious plot is to find frozen male mammoths and extract their sperm then use it to fertilised a female Asian elephant (*Elephas maximus*), the mammoth's closest living relation. The resulting hybrids will be crossed until a 98 percent true mammoth is created.

He hopes to open 'Pleistocene Park' in a 100 square mile wildlife reserve in Yakutia, Siberia. He hopes mammoths will be roaming the area again within twenty years. He also hopes to re-engineer the ancestor of the Siberian tiger (*Panthera tigra altaica*) and prehistoric horses.

He is not alone in his endeavours. Bernard Buigues, the leader of a French expedition, said his team had located a 'frozen zoo' under Siberia's permafrost and knew of a dozen more locations like it. He is searching not only for mammoths but woolly rhino (*Celodonta antiquatus*), the giant elk (*Megaloceros giganteus*) and the cave lion (*Panthera leo spleaus*). Genetic material will be used to try and recreate the animals. Professor Larry Agenbroad, of North Arizona University, principle scientific advisor to the expedition, said:

'There is no significant difference between restoring prehistoric animals and restoring modern creatures such as grizzly bear and bison.'

The mention of bears brings us nicely to our

next Siberian monster. The largest known bear in Russia is the Kamchatka brown bear (*Ursus arctos middendorffi*) but in 1936 Dr Sten Bergman noted that an even larger species may roam the area. In 1920 he had been shown the pelt of one of these giants. He was also shown a skull larger than that of any known bear and a massive paw print fifteen inches long by ten inches wide.

The print had been photographed by Swedish scientist Rene Malasie who lived in Kamchatka for nine years. Could this be a surviving form of the cave bear (*Ursus actos spleaus*)?

More recently an even bigger bear has been reported from the same place. In the 1980s Soviet scientists investigated reports of a titanic white bear said to weigh a ton! Known as the *irkuiem* the monster has a small head, long limbs, and a narrow body. It apparently stalked its prey in an odd fashion by putting its forelegs first then pulling up its hind quarters. The movement is said to be like that of a giant caterpillar.

Reports first appeared in the Russian newspaper *Pravda*. Rodin Sivolobov has collected many eye witness accounts and that rarest of cryptozoological prizes, hard evidence. In 1987 he obtained from reindeer hunters what looked like the pelt of a polar bear (*Ursus maritimeus*), but it was far larger. He reported having shot a massive bear in the 1980s. It was much larger than any other bear he had seen and had a comparatively small head for its body size. He saved the jet-black pelt.

He sent details of the monster bear to Dr Nikolaj Vereshchagin who postulated that the irkuiem could be a surviving form of the short-faced bear (*Arctodus simus*) that like the cave bear is believed to have been extinct for 10,000 years.

The short faced bear, sometimes called the bulldog bear, had a short muzzle and longer limbs than other bears. It would have run much faster and been far larger than any other bear. This animal must have been one of the most formidable terrestrial predators since *Tyrannosaurus rex*. Vereshchagin announced to *Pravda* that he believed the short faced bear still existed, and constituted the eighth species of bear known to man. It would take a bold cryptozoologist indeed to hunt for such a beast, but if anyone has the money I'm your man!

THE SEA APE AND THE SEA COW

Between Kamchatka and Alaska lie a remote chain of godforsaken islands known as the Aleutians. The islands were discovered by arctic explorers George Wilhelm Steller and Vitus Bering on their ship the *St Peter*. On November 28th 1741 even further north than the Aleutians, they were shipwrecked on an Island that was to become Bering's grave and that still bares his name Bering Island. A remarkable naturalist, Steller kept a meticulous journal of his zoological and botanical discoveries both before the shipwreck and after, until his final escape from the Island. Every one of his discoveries was later verified and confirmed by other scientists who visited the area…with the exception of just one, the sea ape.

Before the shipwreck the vessel had been exploring other islands in the Aleutian range further south. On August 10th they were heading for Kamchatka and were hemmed in by wooded islands, when a bizarre animal put in an appearance.

'It was about two Russian ells (five feet) in length; the head was like a dog's, with pointed erect ears. From the upper and lower lips on both sides whiskers hung down which made it look almost like a Chinaman. The eyes were large; the body was longish, round and thick, tapering gradually towards the tail. The skin seemed thickly covered with hair, of a grey colour on the back, but reddish white on the belly; in the water, however, the whole animal appeared entirely reddish and cow coloured. The tail was divided into two fins, of which the upper, as in the case of sharks, was twice as large as the lower. Nothing struck me as more surprising than the fact that neither forefeet as in the marine amphibians nor, in their stead, fins were seen.'
The crew watched this curious animal for two hours as it cavorted around the ship. It came so close that it could have been touched by a pole. It was observed to play with and even eat seaweed. Twice Steller tried to 'obtain a specimen' by shooting at the poor creature. Luckily both shots missed but Steller's rash actions frightened the animal away. He later christened it a sea-ape (on account of its monkey-like shenanigans).

No such animal has been reported before or since but Steller's care as an observer is in no doubt. So what was the sea ape? In some ways it resembles a pinniped (a seal or sealion) but the lack of forelimbs and odd tail flukes show that it is something quite different. All known marine mammals have flippers jutting out to the side not held vertically as this creature did. Perhaps Steller had stumbled upon one of the last of a dying species who became extinct before we could record them. Maybe the sea ape was one of a lost lineage of aquatic mammals more advanced than pinepeds but less advanced than cetaceans (whales and dolphins). We shall probably never know.

Whilst cast upon Bering Island, Stellar made another amazing discovery, but this one was confirmed in a bloody and tragic manner. He came upon what looked like a group of upturned boats in a bay on the Island. He soon discovered that these mysterious giants were titanic sirenians - a group of marine mammals that includes the manatee and the dugong. Totally aquatic and herbivorous sirenians are most closely related to elephants. The new species was a giant of the group at 35 feet in length and tipping the scales at 13 tons or more. They were also the only members of the group to be found in cold water.

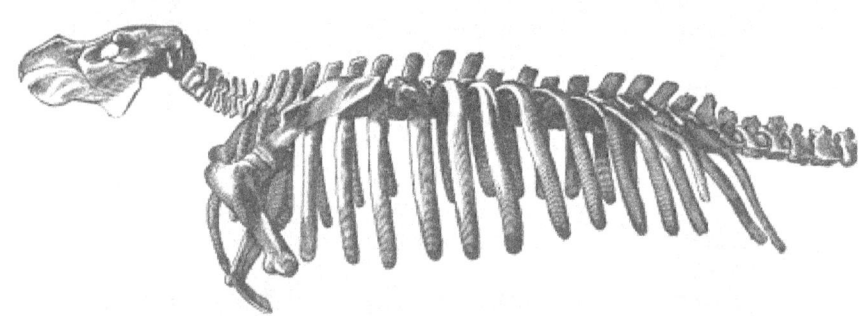

The species was named *Rhytina srelleri,* Steller's sea cow.

Steller innocently condemned the sea cow to death. Such a bounty of flesh,

oil, blubber, and leather soon attracted the greedy eyes of humans of an infinitely lower form than Steller and Bering. The defenceless giants were so intensively hunted that by 1768 it was declared extinct. This assumption may have been in error. Indeed the populations around Bering Island seemed to have vanished. But what if some survivors had moved on elsewhere or the population was not just confined to the Bering Island area. Could some still inhabit the Aleutians?

A E Nordenskjold visited the area on his 1878-1880 polar exploration voyage. On the basis of interviews he believed that the sea cow had lingered until at least 1854. One native, half Russian half Aleutian said that his father, who came to Bering Island in 1777 from Volhynia, reported sea cows still being killed between 1779 – 1780. The hearts were eaten and the hide used to make boats.

Two other natives, Feodor Merchenin and Stepnoff (who is never given a first name in the records), told Nordenskjold that, twenty-five years previously, they had seem a huge sea animal they could not identify on the east coast of the Island.. Their description matched that of a sea cow.

Aleutian natives had indeed reported the creatures, particularly from Attu to the coast of Kamchatka. Peter Simon Pallas, the German Professor of Natural History led a scientific expedition to the Aleutians in 1777 during the reign of Catherine the Great. He discusses various islands in the chain including Attu, Stemiya, and Semichi. He records the people and animals on and around the islands. He says that plenty of sea cows are to be found about all of the islands. Later he goes on to say that the sea cow is not found in the eastern Aleutians such as Kodiak and Umiak (closer to the North American coast).

Russian scientists A A Berzin, E A Tikhomirov, and V I Troinin published a report of a whole herd of sea cows in the journal *Priroda* in 1963. The year previously the whaler *Buran* was in the vicinity of Cape Navarin, south of the Gulf of Anadry. In early July the crew came across a group of dark coloured animals some 6 to 8 metres long. The herd were in a lagoon containing sea cabbage and other vegetation favoured by sea cows. They had bilobate tails, heads that were differentiated from their bodies, and hare lips, all features possessed by sea cows. The animals submerged and surfaced regularly as they swam slowly around. The crew, who knew most sea creatures, had never seen their like before.

The latest sighting of the sea cow to filter through to the west was in 1977. It was originally printed in the Russian newspaper *Kamchatsky Komsomolets.* Ivan Nikiforovich Chelulin, a projectionist for the Karaginskaya culture and propaganda team, was taking part in some salmon fishing operations in Anapkinskaya Bay in 1976. All the local inhabitants took part and formed a collective farm. After a heavy storm Chelulin noticed a strange animal on the tidal belt. It bore a whale-like tail, dark skin, flippers and a head with a long snout. It was different to any other marine animal he or the others had ever seen. When later interviewed by

Vladimir Malukovich, the senior scientific worker for the Kamchatka Museum of Local Lore, he identified a picture of the sea cow from Reconstructions. He was amazed to hear that it was supposed to be extinct.

The museum was set up in 1966 at an Ust-Pakhanchinsakaya school. It holds the remains of several sea cows. The bones of one specimen that was examined by scientists appeared to be from a sea cow that had died as recently as 1967. As with our other Siberian cryptids the area in which they are said to live is remote and pitifully explored and the sea cow has another advantage; it is marine. It seems far from impossible that the largest of the sirenians still inhabits the waters of the western Aleutians.

DRAGONS AND SERPENTS

Giant reptiles seem the last creatures you would expect to be associated with the frozen wastes of Siberia. However, some of the most dramatic accounts of encounters with latter day dragons have taken place here.

The most celebrated case was reported in the November 21st 1964 issue of the newspaper *Komsomol'skaya Pravada*.

Moscow University mounted a geological expedition to explore the mineral deposits of the Kular Range and the surrounding districts. The expedition lasted from June to October and was headed by A.Kharchenkov, an engineer, V.Gomoharov, a post-graduate scientist, and six others.

The team heard rumours of a monster inhabiting Lake Khaiyr. The lake is not large, being some 600 metres by 500 metres. It is, however, connected via swamps to many other small lakes in the basin of the River Omoleya. It is an area of recent disruption of the Earth's crust and is a thermal lake, freezing later than its peers.

After trekking for miles over the freezing tundra the scientists were told by the inhabitants of the small Khaiyr village that no one dared fish on the lake and that no wildfowl would alight on its surface. The lake's depth had never been measured but it was said to be very deep.

The team's biologist, and member of the Yakut Branch of the Academy of Sciences, Dr Nikolai Gladkika, was the first to see the creature. He had gone to the lake one morning to draw water and saw a massive animal that had crawled out of the lake onto the shore. It had a small snake-like head on a long neck, a large body with four short legs and a dorsal fin running the length of its back, and a long tail. Its scaly body was a bluish-black in colour. It appeared to be browsing on the grass, a strange diet for such a creature.

Gladkika ran back to his team mates but when they returned with cameras the monster was gone. They found a large area of flattered grass but no indication that any had been eaten. Perhaps the creature had just been sniffing around in the vegetation. Indeed we shall see evidence later that these beasts are carnivorous in nature.

Gladkika produced a drawing of the animal that he had seen. His illustration
could have come straight out of a medieval bestiary.

The creature strongly resembles a dragon with its slit-eyes, finned back, scaly skin, and snake-like neck.

Fortunately the dragon put in a second appearance. This time, the expedition leader and two members of the biology group were looking across the lake when the monster broke the surface. Its head and dorsal fin were clearly visible as was the long tail that lashed the water, sending waves across the lake.

The expedition leader later postulated that it could be some kind of prehistoric reptile. He intended to return with his team the following year and set up base beside the lake. If he ever did, no reports ever reached the West of a second expedition.

This was not the first time the dragons of Lake Khaiyr had been observed by non-locals. In 1942 two pilots surveying the lake reported seeing two huge animals in the water. They likened them to giant newts with long crests upon their backs.

Russian geologists seem to have luck in monster hunting. In July 1953 a prospecting party led

---

Sadly, new evidence appeared in 2007/8 and as Dr Darren Naish wrote on his blog:

"The entire episode seems to have been a fabrication. As explained in an article published last year in *Komsomolskaya Pravda*, Gladkikh never was a biologist nor in fact even a scientist; he was instead a 'migrant worker' hired to help with the expedition. And while cryptozoologists have often expressed frustration about the lack of subsequent investigation of Lake Khaiyr and its possible monster, it turns out that such investigation actually did take place. Claims that the lake was devoid of fish, that birds wouldn't land on it, and that the locals had a tradition of sightings, were all false. Even better, Gladkikh admitted that he had fabricated the entire event (either 'to entertain himself and his friends or as an excuse for shirking his duties at work', said *Pravda*). "

by V.A Tverdokhelbov travelled to the Sorongnakh Plateau. The survey team arrived at Lake Vorota on a bright sunny morning. Out on the smooth surface of the lake, Tverdokhelbov, and his assistant, Boris Bashkator, observed an object some 300 metres out in the lake.

At first they thought it was a floating oil drum. They soon realised that this was not the case, as the object began to swim closer to the shore. The pair climbed a cliff to get a better view. In Tverdokhelbov's own words:

'The animal came closer, and it was possible to see those parts of it that emerged from the water. The breadth of the foreparts of the creature's torso, evidently the head, was as much as two metres. The eyes were set wide apart. The body was approximately 10 metres. It was enormous and of a dark grey colour. On the sides of the head could be seen two light-coloured patches. On its back was sticking up, to a height of half a metre or so, what seemed to be a kind of dorsal fin which was narrow and bent backwards. The animal was moving itself forward in leaps, its upper part appearing at times above the water and then disappearing. When at a distance of 100 metres from the shore it stopped; then it began to beat the water vigorously, raising a cascade of spray; then it plunged out of sight.'

Was this beast of the same species that haunts Lake Khaiyr? The dorsal fin. and the water thrashing behaviour are alike but no mention is made of a long neck. Perhaps what the witnesses took to be the head was merely part of the bulky body. Only further expeditions can answer that question. Lake Labynkyr lies on the same plateau. It is a big lake at nine miles long and 800 feet deep. It also it has an evil reputation. Local tribespeople are convinced that a 'devil' inhabits the lake. The monster has eaten gun-dogs who have dived into the lake to retrieve shot ducks. One man told of how the brute pursued his raft. He described it as having a huge mouth and being dark grey in colour. Some reindeer hunters observed the monster coil up out of the lake to devour a passing bird.

In 1963 a small expedition visited both of these lakes. Four members observed an object in Lake Labynkyr at about 800 metres distance. It emerged and submerged several times. They could not take a photograph as the sun was setting.

The following year three teams - each replacing the other in shifts - visited both lakes. The third and final group saw the Lake Labynkr monster in the latter half of August.

Two expedition members saw a row of three humps appear 100 metres from the shore. They ran after the humps trying (unsuccessfully) to photograph them. The humps dived and rose together. It was not clear if they were three separate animals or part of one creature.

In 1964 two journalists from the Italian magazine *Epoca* visited Lake Labynkr whilst travelling to Oymyakon. They were told that some time ago a party of men saw a reindeer swim into the lake. The deer vanished and did not resurface. Then a dog swam in and vanished as well. Suddenly and shrouded in mist a vast black monster rose from the lake and snorted. One of the observers, apparently a scholar, was convinced the beast was a dinosaur. The locals flatly refused to take the journalists out onto the lake.

Another story concerns a hunter's dog that swam out into the lake and was eaten by the beast. The grieving hunter constructed a raft of reindeer skin and filled it with hot coals. He floated the smouldering raft out onto the lake and the creature snatched it and dived. The monster reappeared a short time later making terrible sounds.

Russian explorer and writer Alexander Remple has been told many stories of such creatures by natives of the taiga. Known as *paymurs,* they are described as having crocodile-like bodies and heads like sheat-fish, better known as the wels or European catfish (*Silurus glanis*). It should be noted here that Oriental dragons were often depicted as bearing catfish-like barbels on their mouths. One man, Anatoly Komandigu, told of three hunters, who made camp next to a snow-covered mound in the twilight, and lit a fire. The sat with their backs to the mound and warmed themselves as the fire flared up. Suddenly they felt the mound at their backs heave. As they spun round they saw that the 'mound' was a huge reptile covered with thick grey and black scales. It had short legs and a long tail. Needless to say the men fled. Three days later they returned for their equipment. They discovered the remains of an animal, possibly the dragon's prey in the area.

Remple has also been told of another giant reptile in the taiga but one of a different stripe. In 1991 he interviewed Vladimir Semyonovich Kuznetsov who was then 71. He had been a seasoned hunter on the taiga as a young man between the world wars. One night he stumbled upon a hidden settlement in the woods.

Creeping closer with a hunter's stealth, he saw a bonfire in a clearing and heard singing. He saw a semicircle of people around the fire who sang unfamiliar songs. As they sang they bowed reverently towards the setting sun. Suddenly from the direction of the bowing something massive crawled into the clearing. Kuznetsov realised to his horror it was a gigantic black snake some 10 metres long. As the 'snake' reared up he saw it had small fore-limbs and could not have been a true snake. The people, who appeared to be worshipping the monster, began to sing louder in guttural voices.

Overwhelmed with fear, Kuznetsov fled madly into the forest. He lost the trail, and did not know how long he had been running, but his hands and face were covered in scratches. He spent a fearful night up a large pine tree. It appears that he had stumbled on a ritual of some form of dragon-worship cult. If such a thing sounds unlikely, even in the wilds of eastern Siberia, it is sobering to remember that, according to a prominent Tyneside historian, Scandinavian sailors were making human sacrifices to a sea dragon off the coast of northeastern England up until 1928.

A newspaper in Primorije published a story of another hunter who saw a giant snake in the 1940s just after WW2. Whilst hunting in the vicinity of Khuntamy Lake, in an oakwood thicket, he came across a dark coloured snake 10 metres long resting in the branches. The man fired his Berdan rifle at the snake. The enraged animal began to thrash about shattering branches in a terrifying display of strength. The man ran for help but when he returned with his sons and some heavy duty rifles, the snake had gone.

G.E Ribalko saw a snake, five metres long, with a diameter of 10 centimetres in 1978. This was close to the Angu river in the northern reaches of Primorjie. Close by in 1983 Alexander Vodyanin and his co-workers sighted a 10 metre snake whilst hay mowing. In 1984 a black snake of vast dimensions was seen slithering across the road in front of a bus full of coal miners from the Nikolayevsky Mines. It was so huge the miners at first thought it was a fallen log.

More recently the Russian news agency TASS reported that a giant snake six to seven metres long, green in colour and with a sheep like head had been seen by dozens of people in a lake near Sharipovo, South Ural, in Siberia. The monster, with a tree trunk girth, left tracks in the grass on the shoreline and was said to have been photographed. It was believed to have eaten all the fish and frogs in the lake. TASS also reported that some of the older villagers recalled that a prehistoric fish - previously thought to have been extinct - had been caught in the lake some fifty years ago.

The pictures, to my knowledge, have never reached the west, and I have never heard of the 'prehistoric fish' before. Perhaps this is a case of Chinese (or Siberian) whispers and the story was mistranslated.

Could such giant cold-blooded reptiles exist in a cold climate? If so, what are they? We know that dinosaurs could cope with fairly cold climates, but there is absolutely no evidence of any dinosaurs (apart from birds) surviving the mass extinctions of 65 million years ago.

The huge leatherback turtle or luth (*Dermochelys coriacea*) often strays into cold water. It copes with this by being gigantothermic. Its size helps it retain heat. However the turtle is not an elongate animal like our Siberian dragons. Elongate creatures make much less efficient gigantotherms than more stubby shaped creatures, like the leatherback turtle.

Perhaps they hibernate and are active only during the brief summer months. No one will ever know for sure until a dedicated expedition travels to eastern Siberia to search for these saurians.

THE RUSSIAN RODS

We will round off our tour of the Siberian crypto-zoo with the strangest inhabitant of them all. U.S. researcher Jose Escamilla has opened up a whole new field of cryptozoology with his study of creatures he terms 'rods'. These are cylindrical, translucent creatures bearing a long cuttlefish-like fin down each side. They have been filmed many times all over the world, usually by accident. They appear as a blur on the screen but when the film is examined frame by frame the rods are revealed. Escameria believes them to be flying invertebrates of an unknown class, an idea that harkens back to Trevor Constables theories on 'sky beasts'.

[Publisher's Note: See Dr Karl Shuker's paper in the *1996 CFZ Yearbook*, available from editorial address]

The following account was recorded by Koga Soyanka, a native Udege journalist, and was related to him by his grandfather Onenko Tchuimbu.

When hunting in the days before WW1 Tchuibu and his comrades often saw creatures they referred to as 'vidra' (otter). This seems only a reference to the cylindrical shape as the things had on real resemblance to that animal. They were jelly like, flame coloured, with a silvery hue, and flew at a height of six feet.

They were two to three times larger than a real otter and moved about as fast as a man could walk. If chased and struck with a pole they would brake into several pieces that would rejoin in the air. The pole would have an oily residue upon it. The things would often submerge in rivers and reappear on the opposite bank. Most modern rods move at amazing speeds; perhaps these were ill or close to death. We shall never know.

Siberia is a cryptozoological fruit ripe for the picking. Largely overlooked by the outside world, its marvels have lingered un-plumbed by hunters seeking scientific treasure. I have the will - all I need now is the means.

Source material on the 2002 wave of animal mutilations in Argentina, together with an overview of the episodes in question

by

Jonathan Downes

(AUTHOR'S NOTE: For reasons of contextual accuracy no attempt has been made to anglicise any of the translated news items sent to us. Spelling mistakes have been corrected but no change has been made in the grammatical construction of these reports)

During the summer months of 2002 Argentina underwent a wave of animal mutilations almost unparalleled in our experience. I have written elsewhere (Downes J and Wright N., *The Rising of the Moon* Domra, Corby, 1999; Downes. J., *Only Fools and Goatsuckers* CFZ, Exeter 2001), about my theories concerning the links between waves of animal mutilations and socio-political and religious unrest. It is interesting, therefore, to note that at the time these mutilations were occurring, Argentina was going through an unprecedented wave of socio-political chaos. It is also interesting to note that, in my opinion at least, this wave of animal mutilation activity can only be compared realistically with the outbreak of chupacabra-related activity, which took place in Mexico during 1995/6. This is related at some length in Dr Rafael A Lara-Palmeros's article on the outbreak of mutilations in Mexico, which was published in the 1997 edition of this journal.

When starting to compile this paper, I was torn between two concepts. Firstly I was tempted to sort the news items in my burgeoning files on the subject into themes, but I thought better of it. I have written elsewhere about the anatomy of animal mutilation episodes and, as will be seen in this paper, the events in Argentina during the summer of 2002 followed the pattern that we described in *The Rising of the Moon* and in *Only fools and Goatsuckers* to a remarkable degree. I therefore decided, after much soul searching, that, in this instance, it would be more appropriate to present the source material in roughly chronological order. In *The Rising of the Moon* we describe how, during the UFO wave which covered East Devon for a period of several months during the summer of 1997, the researchers at the CFZ felt that they were becoming 'shell-shocked' at the sheer volume of information that they were receiving on a daily, and sometimes an hourly basis.

Although the events in Argentina did not have the same effect on the CFZ researchers, taking place as they did at the other side of the world, the sheer volume of information that was coming in, sometimes several times a day, produced a similar (though much diluted) reaction, and one can only imagine at the effect that these events would have had on a researcher in the field such as veteran UFOlogist and Fortean Scott Corales, to whom we are indebted for the vast majority of the source material that we are placing here before you.

The earliest reports were posted in mid-June:

Source: Diaro Rio Negro Online (Argentina)
Date: June 15, 2001

None of them presented gunshot wounds or piercing incisions. There is considerable fear and doubt among the population. "Intelligence played a role in this," said one veterinarian. There are too many questions, and mystery continues to enshroud the discovery of mutilated cows in the province of La Pampa. Up until yesterday, the dead animals missing certain parts of their bodies, such as tongues, udders, anuses and eyes, had been tallied at 20. It was possible to determine that the animals died painlessly, apparently without suffering, and none of them presented signs of having been shot nor stabbed, although this does not clarify the real cause of these deaths. Furthermore, under no circumstances did the cows bleed to death. Given the evidence on site, the cattle offered no resistance - had they done so, tracks would have been found and some muscular contraction would have been evident [in the carcasses]. But that did not occur in these cases. Experts are certain that the animals died in the very same field in which they were found and could have hardly been dragged there, since the lightest of them weighed 350 kilograms and secondly, because no vehicle tracks were found on site. Yesterday, all eyes were on the Pampan veterinarians working on the case, who were nevertheless unable to come up with an explanation for many of the aspects surrounding these mysterious deaths. One of them opined that within this field, there does not exist the technology to perform the almost perfect work evinced in the cattle mutilations. As this newspaper was able to ascertain, both in La Adela (where the latest findings occurred) as well as in other communities, two opposite perspectives have come up: many people fear the unknown while others took the matter lightly. The fact remains that 20 cows are now dead.

Yesterday, José Casiavillani, a veterinarian from the municipality of La Adela, who performed the necropsy on the cows, was interviewed by Radio Manantial's "Antes que Nada" program. The following is the dialogue between the interviewer and the professional:

Interviewer: "The subject has no explanation to us, the laymen, but does it also pose a mystery to you, the expert?
Veterinarian: "That is the case, at least up to now. The field I surveyed yesterday and the day before contains the most numerous cases, there were 11 dead animals and not 10, as originally believed. They were all in the same field."
Interviewer: "How long ago did these deaths occur?"
Veterinarian: "Possibly Saturday or Sunday. It turns out that the claim was only made on Tuesday and the visit was made on Wednesday. I travelled approximately 5 kilometres, starting at the farmhouse and plunging into the countryside, and the cases begin to appear, all of them following the same pattern, which we could define as a circle, if it were possible to see it from above."
Interviewer: "Arranged in a circle?"
Veterinarian: "Yes. It begins some 50 meters from the farmhouse with four dead animals." Interviewer: "50 meters away from an occupied house?"
Veterinarian: "Yes, but at that time the person in charge of the property wasn't in. So it was that when he returned, he was confronted by this scene, got on his horse, toured the field and came across more dead animals. These cases are very particular, almost inexplicable, because they show a kind of cauterised cut. No hemorrhage occurs when the incision is made.
Interviewer: "You, sir, are a veterinarian. Is it possible to do it as it was done?"
Veterinarian: "Well, it's possible, but not with field elements. It involves technology, such as an electric scalpel that cuts and cauterises at the same time.
Interviewer: "Was there an abundance of blood?
Veterinarian: "No, not in the cut itself, and that drew my attention. All of the animals are missing an eye, for example - the one that's clearly visible, because the animal was left lying on its side." Interviewer: "How were the cows slain?"
Veterinarian: "They don't have gunshot wounds or signs of piercing or perforation."
Interviewer: "So what could have caused their deaths?"
Veterinarian: "That's the unanswered question. The animals' deaths is sudden. One knows when an animal is agonizing because it kicks around and leaves traces on the ground. It was the first thing we veterinarians notice when we visit a field to determine how an animal was slain."
Interviewer: "They died where they were found?"
Veterinarian: "Evidently so. There are no tracks of any kind - human, animal or vehicular."
Interviewer: "What do these animals weigh?"
Veterinarian: From 350 kg upward. These are cows which could not have been moved in any way. There are no bulls among the slain."
Interviewer: "Were the udders severed?"
Veterinarian: "Yes, in a very dramatic way. Cows have four nipples and four mammary glands. We found udders in which a circular cut around the nipple was made. Only one nipple and the gland corresponding to it were extracted, but the rest was untouched. The farm hands who surveyed the field, and are good at the task, were unable to find any tracks. Cows, while not given to attacking and considered harmless to everyone, tend to flee when they see or hear something strange. In these cases, fear overwhelms their curiosity and they take off.
Interviewer: "Do cows defend themselves when attacked?"
Veterinarian: "Yes, and that's another mystery. That's why I said that there is intelligence at work here, because, how does one reach the site of the latest events without leaving traces. This field is in a dirt road area, and from that route to the field there is a five kilometer distance, and then a circular route of another five kilometers. It is very hard for one or several persons to catch a cow in the wild. Cows are very slippery; if they hear or see something strange, they tend to flee."
Interviewer: "Was there any blood in the field?"
Veterinarian: "No, and unlike other cases, there was blood in the animal - in the heart and

throughout the circulatory system."
Interviewer: "How long was the watchman away from the house?"
Veterinarian: "He was on the farm up 'til Friday, and spent Saturday and Sunday in town. He had never seen anything that made him feel suspicious. The watchman could not imagine the scope of what was happening, and when he came across one of the animals, began to butcher it to feed the dogs, but the dogs refused to eat the meat. None of the dead animals was eaten by other carrion beasts."
Interviewer: "Are you afraid?"
Veterinarian: "No, not at all."

The fact is that this newspaper ascertained yesterday that surprise and fear of the unknown was in evidence throughout the region.

This first report is presented in a matter of fact way. At the CFZ we are presented with dozens of animal mutilation reports throughout each year. Although this one is more in depth than most there was nothing about it to suggest that it was the first incident in what was going to prove to be perhaps the biggest and most important animal mutilation episode yet recorded. The final lines are however somewhat portentous in view of what was to come. They also suggest that these were by no means the first animal mutilation cases to be reported from rural Argentina.

The accounts of the injuries "the dead animals missing certain parts of their bodies, such as tongues, udders, anuses and eyes" and the fact that there was no discernable cause of death are highly reminiscent of all the major outbreaks of cattle mutilation which have been reported over the years.

Over the next few days the reports continued to come in:

Source: Diario 'La Nueva Provincia' (Bahia Blanca - Argentina)
Data 6-16-02   New Mutilations in Rivera and PringlesStrange Mutilation Wave Continues

The mutilation in the [town of] Rivera is the first one to involve a horse. The animal had died from natural causes and its owner transferred it to a remote area. Days later, the carcass showed signs of strange incisions. Meanwhile, the first cattle mutilation in [the town of] Coronel Pringles was registered yesterday, involving a 220 kg. steer. Veterinarian Marcelo Erro and stallholder Abelardo Vivas claim having never seen anything like it.

CARHUE and CORONEL PRINGLES (from our offices) - Two new animal mutilation cases shook the communities of Rivera and Coronel Pringles.

In the first community, belonging to the Adolfo Alsina district, residents had still not overcome the surprise and uncertainty caused by the appearance of a bovine found mutilated at a farm, when word was received that a horse had suffered similar injuries. "A psychosis has been unleashed throughout the residents," believes veterinarian Jorge Robles, alarmed due to the fact that theories over the causes of the mutilations have spread like wildfire in Rivera. "The horse died of old age some 20 days ago, right near the house. For that reason I decided to rope it by the legs and drag it to a secluded area. I went by many times in recent days but never paid the carcass any mind until I saw what had happened and was startled," related the horse's owner, who declined to identify himself.

"The animal's missing an eye, like the one that turned up in Salliquel', and I was startled by the cut that can be seen around its teeth and below its jaw, which is a perfect cut," added the cattle rancher. He added that the horse was also missing its tongue, despite the fact that its jaw was clenched shut. "It's weird, because when an animal dies it stiffens so much that it's hard to open its mouth."

The cattleman also noticed that the horse was missing its anus and genitals, as well as skin from the inner section of the genital region. "Under its tail, it had been clipped perfectly down to the hair, but with the abdominal wall showing," he described.

The rural location where the mutilated horse was discovered is west of Rivera, some 12 km away from the town.

The horse was inspected by veterinarian Jorge Robles, who also analysed the dead cow found in Arano. "The incisions are the same. The horse is missing its tongue, an eye, an ear and its genitals." Robles recalled the autopsy he performed on the cow he found at Arano, discovering that its internal organs were intact.

"But when I reached the pelvic cavity, I found a hole resembling a tunnel, and I couldn't find the uterus or the ovaries. They also took its eye and ear, and it was missing a patch of skin some two or three centimeters wide, bordering its lips." He further added that the cow had lost the moving part of its tongue, although without having made any incisions in its neck area. The incisions on the horse are identical. The veterinarian said that in both cases it is impossible to determine an exact pathology. Nor can it be considered, he added, that "classic surgery was employed in making the incisions."

"The wounds are not burned, and while my understanding of laser surgery is scant, I found that there is no laser that can cut hide in that way. Furthermore, there is no sign that the cutting element has touched anything other than hide," he concluded.

The vet dismissed the possibility that a wild animal could make those incisions. "The bird or animal that attacks a carcass isn't after the eye, the ear or the mammary glands. They are carnivores and feed on flesh. In this case, 80% of what's missing is hide. This is what fills one with doubts." The vet confirmed that the places where the animals are found are removed from trails or roads. "The people who discovered the first cases noted that all of the carcasses have their heads toward the east and their tails toward the waist. Since it isn't easy to find the cardinal points in a field, I took a compass with me when I went to see that cow again. To my surprise, the head is pointing due east and the tail due west." "What happened cannot be fit within the framework of medical science." he added.

At this stage the interpretations, which were implied in the newspaper reports, were that the perpetrator of these attacks was human. It wasn't until the next report that the first inferences were being made linking these attacks with the predations of the chupacabra - a creature that is described in greater detail in my *Only Fools and Goatsuckers:*

Source: Diario "Nuevo Día" (Colonel Suárez, Prov. of BuenosAires)
June 15, 2002

Animal Mutilations: Veterinarian Who Examined First Case Believes Culprits Arrived by Air   Doctor Daniel Belot, a SENASA Technician in Salliqueló, doesn't believe in Chupacabras, but has never seen anything like it...   He was the veterinarian who analyzed the first animal mutilation in the area, in Salliqueló. He has examined over a dozen cases for which he can find no explanation, and is disconcerted. He does not believe in the intervention of any agencies other than human, but he discards nothing in the light of the magnitude of what he has observed.

He is convinced that the perpetrators arrived by aerial means and is certain that the animals were slain elsewhere and subsequently dumped in the field. He asks that all strange animal deaths be reported promptly.

Dr. Belot is startled, moved and disoriented by the sudden appearance of mutilated animals throughout the region. In spite of his considerable professional experience, Belot, who is a technician for SENASA, claims having never seen anything similar and believes that only those who have seen any such case can understand the magnitude of the events.

The expert has been asked to analyze over a dozen cases, and common patterns appear in all of them to which no explanation can be found. Although he is unwilling to forecast the events, he is certain that more cases will only confuse him more rather than clarifying matters.

The near-certainty that the perpetrator arrives by air, and the certainty that the deaths have occurred 3 to 4 days earlier at a location other than where the carcasses were found only adds to the state of confusion.

His account of the facts and some of his interpretations merit analysis, since he is an acknowledged professional with a solid reputation throughout the region, aside from being an official for a national delegation of SENASA, with the added responsibility and weight

Belot began his story by recalling that "the cattleman told me he had a dead animal in a field and his attention had been drawn to the fact [the animal] seemed to be skinned to the bone on one side, which I thought was impossible, because it was something I wasn't familiar with, something utterly abnormal. For that reason I told him that some animal must have eaten it, which the cattleman completely rejected, saying that no animal eats in a straight line. Therefore, I resolved to go and see what was going on."

"I was confronted by an unnerving sight," the professional told NUEVO DIA, stressing that "those who haven't seen it cannot understand the magnitude of the situation." Belot explained that "the animal lay on the ground like a hare, and the entire left side of its face was skinned to the bone beneath the eye. All of its molars were visible.

"When we performed the necropsy we found that it was missing its tongue, all of its vocal apparatus, which is to say the larynx and part of the pharynx, and something very odd: there was no blood inside or outside the animal. It was perfectly clean. That came as an enormous surprise for us." He confirmed that there was no tearing [of flesh] of any kind on the animal, therefore discarding the likelihood that predators would have attacked it. "It is a deed that appears to have been carried out by humans, but even so it's something very hard to do." When asked if he was able to come up with any explanation whatsoever, Dr. Belot confessed that "my curiosity has not yet been sated, since I've sent samples to the University of Buenos Aires' School of Pathology and the only response I've received is that how the incisions were made cannot be determined." The professional stated that he did not establish the animal's cause of death, going as far as to state that blood samples taken from other animals to detect strange substances "have yielded no results so far. The ones I have received confirm that there is nothing strange in the [animals'] blood."

Belot, who has analyzed over a dozen of these mutilation cases, remarked that vital organs were missing in most of them. " In the first case, which is the one I'm discussing here, the animal was missing all of its maxillaries, but others were missing testicles and penises, others were missing ears, others were missing mammary glands, and still others were missing rectums and vaginas. All of this suggest some kind of scientific research, but don't ask me why a scientist is going to conduct research in the middle of different farms without asking for permission, because I can't imagine why." Belot reiterated his astonishment as to the absence of tracks around the mutilated animals, wondering how it is possible "for no other animal to come near." When Dr. Daniel Belot reached the first animal he inspected, it had been dead for 3 or 4 days, a situation which has repeated itself in many of the cases he analyzed. "In many cases they were not yet bloated and did not give off any odor. In others, they presented normal alterations, which increases the uncertainty." The cases observed by the SENASA technician occurred in the Salliqueló region, but were later extended to Casbas and Guaminí.

Belot acknowledged that in recent days he learned of the Chupacabras legend. While not dismissing it, he is not a believer. "The facts occurred, they are very strange and cannot be disputed, but I don't know what to attribute them to. I wouldn't want to chance it." The professional does not recall having seen perfectly circular burned grass spots or any alterations near the mutilated animals, as other experts have stated, explaining that "the only mark I saw around the animals within a ten meter radius, is that the grass has continued to grow [...] but no animal will step into that circle, despite the fact that the grass is in good shape." Two months after the first discovery, Belot points out that "after some time, the dead animals have already been attacked by predators. But until recently, they wouldn't come close." Regarding the behavior of the herd's other members, the veterinarian noted that "they appear indifferent, looking from a distance without coming close. What is commonly seen in the countryside is that the rest of the animals sort of "mourn" the deceased animal, and that has not occurred in this case."

The professional further emphasized that "the animals were killed elsewhere and dumped there, and that indication I'm inferring from the way in which the carcasses have been positioned."

Translation (C) 2002. Scott Corrales, Institute of Hispanic Ufology. Special thanks to Gloria Coluchi.

Then came the beginnings of what many perceived to be a cover up. As discussed in *The Rising of the Moon* allegations of Government involvement in episodes of Fortean phenomena appear to be an integral part of the phenomenon itself. Even during our investigations into the Exmouth UFO flap of 1997 we encountered claims that shady branches of officialdom knew perfectly well what was happening and were, if not directly responsible for it, colluding in attempts to keep "the truth" from the public at large.

A facet of these episodes which is less well known but which is becoming ever more apparent is that whenever there is a spate of incidents such as the ones in East Devon or these incidents in Argentina, officialdom tends to come up with an explanation bordering on the farcical. Whether making ludicrous excuses is an integral part of Civil Servicedom across the world or whether these are just the machinations of the Cosmic Joker underlining his efforts to make the omniverse a weird, surreal and ultimately ludicrous place, it is not the purpose of this paper to speculate. However, in our experience there have been very few official explanations, which are as ridiculous as this one…

Reuters. Posted July 1 2002.

BUENOS AIRES, Argentina - Recent mutilations of cattle and horses in the Argentine countryside were the work of rodents, scientists said on Monday, not ritualistic slayings by extraterrestrials or vampires as some farmers feared. Argentina's national food and animal health inspection service Senasa sent its own "X-Files" scientists to the remote plains to look into the deaths of farm animals found mutilated and drained of blood. Frightened farmers claimed to have seen bright lights and UFOs in the area where the deaths occurred.

But Senasa officials said the dozens of livestock whose genitals, tongues and other organs appeared to have been removed with surgical precision were victims of rodents, foxes or other animals. Senasa said the farm animals likely died from common infections and wild animals later mutilated the corpses. "We see these as natural deaths (and) there is clear evidence of the presence of rodents and birds which led us to our conclusions," Senasa President Bernardo Cane told reporters.

The strange circumstances surrounding the deaths - one horse's hoof had a circle drawn into it and some animals were surrounded by charred grass -led some locals to insist the deaths were the work of little green men, vampires or a satanic cult. Senasa gave no explanation for the burned grass and the circle on the hoof. Senasa said its specialists conferred with scientists in Texas who had investigated similar cases in the 1970's and arrived at a similar conclusion.

By this stage many of the news reports referred to an ongoing wave and gave the impression that it was a reworking of a Senasa document intended to dispel some of the fear and trepidation that had begun to sweep the rural communities. As can be seen from the next report, the reporters start to refer to the victims as "Bovines" or "Equines" rather than as cattle and horses in an attempt to de-anthropomorphise the attacks. The fact that SENASA had come up with such a ludicrous explanation as the one about the mice is also proof that by this stage a very real panic and terror was beginning to grip the rural communities.

SOURCE: El Diario de la Republica
DATE : July 9, 2002

ANOTHER MUTILATED COW IN SAN FRANCISCO SAN LUIS.

Just when the mystery of the cattle mutilations appears to have been dispelled by a scientific explanation, a five month old bovine was found dead in San Francisco del Monte de Oro, showing the same incisions as the others which were a source of intrigue throughout the country. The discovery was made on Friday night at a ranch known as "El Quebrachito", 8 km west of the entrance to the community. According to Ruben Diaz, a journalist with San Francisco Televisión, the person who saw the animal first is a man known as "El Negro" Vazquez. The animal and the field belong to Alberto Martin, former superintendent and a respected businessman in the community. According to Diaz, the bovine is missing several internal organs and its flesh had a strange look. "It's as though a machine had gone into its mouth and sucked out everything inside." The external examination showed that the bovine was missing its ears, tongue, intestines and one eye. Its anus appears to have been extracted and the entire area presents a wound similar to ones made by burns. "Part of the tail has similar lesions," noted the journalist. People who know the area are startled by the fact that dogs did not bark out a warning, nor made any sound whatsoever. They are further certain that the event was discovered shortly after it took place. According to Diaz, on Friday afternoon, the field owners' children were cutting down a tree not far from where the dead cow was found. In an interview granted to Diaz's TV show, Martin stated that he had never before seen anything similar.

One has to assume that the "scientific explanation" referred to in the text is the official one; that the majority of what appear to be post mortem mutilations were carried out by the predation of various small rodent species. This explanation, albeit unlikely, was bolstered up by the production of several unimpressive photographs showing small mice clambering over the bodies of fallen cattle. One of these photographs was published in *Fortean Times* and others were circulated on the Internet.

Within days there were more attacks:

DATE: 07-07-02 ANIMAL MUTILATIONS NEAR RUFINO: THE COLONIA TRES ARBOLES CASE

A mutilated calf was found on a field located 10km SE of Rufino, owned by one Mr. Caunedo and currently leased by Mr. Camilo Lisiardo. Mr. Lisiardo habitually tours the field in order to observe the state of his cattle. On June 28, he noticed that his herd was doing well. When he returned on Monday, July 1, he noticed a dead calf in a rather remote section of the field. Its body presented strange mutilations. IFOR got in touch with Lisardo on July 2 and were immediately taken to the site. On site, they found the carcass of a small calf lying on its left side with its head pointed toward the southeast. They immediately became aware of a clean cut running from its nose to the throat. The lower maxillary and a small portion of the upper maxillary were completely clean, as though the animal had been dead for several months. It was also noticed that its right eye was missing and the orbital cavity was completely empty. Its belly, particularly around the navel, featured a circle almost 4 cm in diameter. Its hindquarters showed signs of predation, but Lisardo, who had seen the animal a day earlier, had detected a well-delineated circle in the anal region with the extirpation of said organ. The presence of the tongue was not detected through the lower maxillary. IFOR returned to the site on Wednesday the 3rd with veterinarian Hernando Brandino, who made a full examination of the animal, stating that the animals hindquarters had been predated by carrion eaters. After careful observations, posing a few hypotheses and making incisions with a knife, he was able to corroborate the similarity between the existing cuts and the ones made. While he did not altogether discard predator action, he leaned toward the evidence that said incisions could be due to the action of a sharp instrument. He was also alerted by how clean the lower maxillary looked, which would be the case in an animal that been dead for a long time. This find, which can be included in the long list of mutilations recorded in different parts of the country, since it follows the same patterns, cannot be linked to the observation of lights or other anomalous phenomena, since these have not been detected.

Regarding this specific case, we respectfully disagree with the explanation provided by SENASA that names the Common Red-Muzzled Mouse (Oxymycterus rufus) as the cause of these mutilations for several reasons: among them, should this rodent exist in our area, its population density would be too low to provoke such depredations. On the other hand, there is no evidence that these rodents can make such straight and perfect incisions.

Detail of the head mutilations.

A straight incision can be seen along the neck and running to the nose. The lower maxillary is completely skinned and the tongue and an eye are missing. In the case involving this mutilated calf, it is very likely that the incisions were made by an edged instrument. What remains to be answered is who or what did it and to what end.

By this stage, researchers such as those at the CFZ studying the phenomenon from afar were beginning to realise that not only was this 'flap' one of major importance, but that it was following the same course as had previous ones. In our book *The Rising of the Moon* Nigel Wright and I pointed out that Fortean phenomena are seldom seen in a vacuum. Where there is a series of reports of phenomena of a specific type (in the 1997 cases, UFO sightings and here animal mutilations), other Fortean phenomena soon follow. What happened next, therefore was no surprise to any of us:

SOURCE: Diario "La Gaceta" de Tucuman
DATE: July 10, 2002 GOATS AND COWS MUTILATED AT THE LEALES INTA
 Resident claims having seen a large light. Animals found in Tucumán are missing soft parts such as eyes, gums and genitals

Tucumán couldn't be left out. Eight bovines and seven goats were found mutilated in recent days in the pasture field owned by INTA at Santa Rosa de Leales. In all cases, the soft tissues were missing, such as eyes, tongue, gums and genitals. Neighbors were surprised to see that the animals were bloodless. For two months now, cases involving mutilated animals have kept Argentineans in suspense. From extraterrestrial beings to the legendary "Chupacabras", theories regarding the authors of the attacks were numerous. Two weeks ago, SENASA issued a report stating that the attacks had been carried out by a predator known as the "red-muzzled mouse", which had undergone a mutation and become carnivorous. In Leales, Maria del Carmen Reyes claimed having seen a bright light, which was followed by the discovery of the dead animals. The mystery continues...

With this, the first of the UFO reports events were becoming too much for anyone, whether researcher or Government Official to ignore. The SENSA claims that the mutilations were all the work of tiny pampas rodents was simply treated with scorn by everyone involved. As the first month of the killings drew to an end there was another spate of incidents:

SOURCE: Diario La Nueva Provincia (Bahia Blanca, Argentina)DATE: Thursday, June 27, 2002   A WILD BOAR IS ADDED TO THE LIST OF COWS, HORSES, SHEEP AND A GUANACO
Wild animals are also experiencing strange mutilations The case is suggestive, because wild boar are highly distrustful and detect - through their sharp olfactory system - the presence of strangers. On this occasion, an eight month old boar was mutilated. This is the second instance of a wild animal being mutilated since the appearance of a mutilated guanaco in Chubut.

While the mutilation count is now reaching 170 and the cases recorded extend to nine Argentinean provinces, a new ingredient has been added to the phenomenon: the discovery in Rio Colorado of a wild boar carcass showing strange incisions. This is the 6th species hit by the phenomenon, since the mutilated remains of cows, horses, sheep, pigs and even a guanaco (in the province of Chubut) have been found.

Cattleman Nestor Soulé, owner of a rural property some 55 km from Rio Colorado, in the Department of Pichimahuida, found the boar specimen last Sunday near a gully. The animal, he explained, had geometrically perfect incisions and was missing the anus, tongue and jaw. To the surprise of the rancher and the farmhand escorting him, the animal was still soft in spite of the -14 C temperature marked by the thermometer. "It's impossible that an animal could remain in this condition after dead and after the cold, but I was surprised not to find any tracks nearby," said Soulé, who is aware of the "map" presented by the animals and the varying forms and conditions of the pasture surface. The rancher chose not to file a police report, but did put the boar carcass in his pickup truck to show his neighbor. He later left it in the wilderness. "Due to its traits, a boar is an animal that is able to smell any living creature a quite a distance, even thought [the boar] may be young....this is why it flees at the slightest sign of any peril," Soulé explained. The rancher thus made clear why this case is so fascinating and different to the ones involving breeding animals. "I'm not a researcher or anything similar, and that's why I don't know what to say, but this is strange and I never saw anything like it before."

In La Pampa

The carcass of an Aberdeen Angus bovine missing parts of its body were found last weekend in a field located 2 km from the town of Embajador Martini, province of La Pampa. The professionals who analyzed the case verified that it was missing part of its tongue and mammary glands, and that its left maxillary had been sectioned off. A mysterious situation also occurred to the north of the Pampan town of Rancul at the "Los Caldenes" ranch, where a cow had died of natural causes in the afternoon only to be found with incisions on its jaw and missing its tongue the next day. The province of Entre Rios has also the epicenter of these events: in a field located 2 km from Rt.11, a rancher found a 170 kg heifer and a 500 kg. milk cow some 800 meters from each other, both presenting the notorious incision marks. **In Carhué** Cases involving strange animal deaths are still a subject for discussion in this city given the fact that mutilations continue to appear day after day for unknown reasons.

The discoveries made by veterinarian Fernando Lopez are complemented by those of his colleague, Horacio Volpe. Last week, the professional was summoned by the owner of a rural property located 25km from Carhue - no personal information mentioned for privacy reasons - where an 8-9 month calf showing the trademark incisions recorded in the area had been found.

The professional noted that his attention was drawn by the perfect incisions, the flaccidness of the animal - "which seemed recently dead," he noted - and the absence of blood. While Volpe preferred not to hazard a theory, he pointed out that an employee of the property told him that his brother, who was cultivating during the night, "saw some intense red lights, as though from a laser, moving very quickly across the countryside." Veterinarian Omar Fernando Lopez also observed a new case in a rural establishment close to this city: it involved a pregnant cow found in the fields of Juan Carlos Robilotte, showing the same incisions, with the difference that the cow was about to give birth. "There was a circle [on the left flank, above the udder] through which half of the calf's body was protruding, but I couldn't ascertain if there were any lesions, since it was in an advanced state of decomposition."

A Racehorse

A racehorse belonging to the Quarter Horse breed, belonging to stables near the Rio Negran town of Choele Choel, was found dead with strange mutilations. "It was as though it had been hollowed out from within," reported police sources in describing the stud horse, of considerable economic value.

Veterinarian Carlos Montobbio, who certified the case, reported that the horse was missing an eye and part of its tongue, but no incisions on the maxillary had been performed. It had also been castrated and a significant part of its small intestine and rectum had been removed. The expert dismissed the possibility of an attack by "yellow-jacket" wasps: "the surgical incisions presented by the animals cannot be made by insects," said Montobbio. **Lights Over Carhué** Several residents claimed to have seen strange lights allegedly related to the presence of UFOs. Manuel Alesso and Raul Blengio, two rural livestockers owning property in the Paraje Cilley area, 10 km from Carhué, saw these lights in the same way as did other witnesses from different parts of Buenos Aires province. "Between 7:30 and 8 p.m., I was touring the fields of Arroyo Venado when I saw two lights to the northeast," said Alesso. He added that he stopped his pickup truck and flashed his headlights at the lights.

"The light began to shine even brighter and seemed to descend, which is why I went over to where Raul (Blengio) was planting and asked him to look. The lights were white, round and gave off a sort of haze," he noted. He said that behind the light there was another less shiny one. Raul Blengio added that the lights moved slowly: "The weren't stars...they were strange moving lights," explained Alesso, while adding that a third witness accompanying them also saw the lights. "I'm certain they have something to do with the cattle mutilations," he added. Carolina Montenegro, a young Carhuensan about to graduate with a degree in Business Administration, spoke of having a similar experience: "On Friday night, around nine, I was in the countryside and to my surprise I saw a small light, orange in color and smaller than a star, and at the height of said lights there were no other stars to be seen. As I watched, the light changed colors from orange to red, then to green and then back to orange until after 5 minutes it became smaller and vanished."

With the death toll having reached in excess of 170 animals, even with the Argentine economy in ruins and riots in the streets it was obvious that the Argentine Government needed to act. There is a well known syndrome within politics at all levels. It goes.

1. A situation is out of control
2. We have to do something
3. This is something
4. Let's do it

In this case, (unless the conspiracy theorists are correct for once) completely ignorant of the true cause of the mutilations, the Argentine Government decided to buy themselves some time by casting doubt on the SENSA report. Despite the fact that the findings were completely ludicrous, the report had been quoted widely. It is tempting to theorise that this next news item shows the Argentine Government desperately trying to throw up a smoke screen by blaming as much as possible of the confusion on SENSA, much in the same way as SENSA had done to the unfortunate field mice:

SOURCE: El Diario de la Pampa
DATE: Tuesday, July 9, 2002 AUTHORITIES STATE THAT THE RED-MUZZLED MOUSE DOES NOT EXIST IN LA PAMPA

Government dismisses SENASA report on cattle mutilations

The provincial government dismissed yesterday the official report presented by the National Health and Agroalimentary Quality Service (SENASA) regarding the causes of death and mutilations of dozens of bovines, since the "red-muzzled mouse", the alleged perpetrator of dozens of bovine deaths and mutilations, is not found in La Pampa.

This was made clear by the Minister of Production, Nestor Alcala, who pointed out that the rodents of this species "are unknown to me, nor do I believe they form part of the Pampan fauna." Veterinarians and agronomist engineers echoed this sentiment. Gustavo Siegenthaler, director of the National Museum of History of La Pampa, noted that "this species has not been found in the surveys we have conducted." From 1986 to 1992, Siegenthaler headed a multidisciplinary team which produced a report entitled "Survey of Vertebrates in the Province of La Pampa."

"We have placed between 70 and 120 traps each night and have never found that species, and it does not appear in the bibliography either," he stated. The book "Mammals of Argentina" by the Migule Lillo Institute, indicates that the "red-muzzled mouse" lives from Mesopotamia [Trans. Note: name given to the region of Argentina between the Paraná and Uruguay Rivers] to northeastern Buenos Aires province. "It cannot have spread to other areas, because it would have been detected," he explained. "And in the event that they were the authors of the mutilations, it would be assumed that they would be more or less significant populations, thus making them easy to find, which has not happened." Veterinarian Maria Parturlane said that "the anapathological lesions (on the mutilated cows) cannot be said to have been caused by a mouse." Medical veterinarian Alberto Pariani, Academic Secretary and Professor of the School of Veterinary Medicine at the National Univ. of La Pampa at Pico, after reading the SENASA report, considered that "there are always field mice, but for example, in the cases we found in the field, there were no traces of rodent fecal matter."

Meanwhile it was learned yesterday that a mutilated cow was found in the "San Juan" pasture field near Algarrobo del Agula, owned by Pablo Bravo. Alkali said that "the report these people have put together (meaning SENASA, based on research from Univ. del Centro in Buenos Aires) may be what they saw, but I don't know that it agrees with what is happening in La Pampa or other parts of the country." The SENASA report states that the mutilated bovines died "due to natural causes" and that subsequent lesions on the hide and organs were caused by predators such as rodents and even foxes in some cases. The health organization thus attempted to pour cold water on the subject, but few have believed in this version of the events. Furthermore, the Uruguayan government ruled, on the same day that the report was issued, that mutilations in that country were produced by the German Wasp (yellow jacket).

The Argentine Government, it should be noted, tried to give their pronouncements a certain gravitas by citing an even more ludicrous report from the Government of Uruguay. In the meantime the events were spiralling out of control with a new UFO sighting and the first inferences that *el chupacabra* is some kind of extra-terrestrial:

SOURCE: Diario "La Gaceta" de Tucuman
DATE: July 10, 2002

GOATS AND COWS MUTILATED AT THE LEALES INTA

**Resident claims having seen a large light. Animals found in Tucumán are missing soft parts such as eyes, gums and genitals

**Tucumán couldn't be left out. Eight bovines and seven goats were found mutilated in recent days in the pasture field owned by INTA at Santa Rosa de Leales. In all cases, the soft tissues were missing, such as eyes, tongue, gums and genitals. Neighbors were surprised to see that the animals were bloodless.

For two months now, cases involving mutilated animals have kept Argentineans in suspense. From extraterrestrial beings to the legendary "Chupacabras", theories regarding the authors of the attacks were numerous. Two weeks ago, SENASA issued a report stating that the attacks had been carried out by a predator known as the "red-muzzled mouse", which had undergone a mutation and become carnivorous. In Leales, Maria del Carmen Reyes claimed having seen a bright light, which was followed by the discovery of the dead animals. The mystery continues

A couple that was driving along last night at around 21:00 hours from the SETIA campground toward the city along the peripheral road witnessed a large, sky-blue light moving over the lagoon, changing color and size as it did so. Once this newspaper was alerted, a reporter from this newsroom was able to see the light, which at the time remain static over Mt. Brown, seen from the San José Beach Facility. Its color was now red and it gave off flashes. Minutes later a local resident contacted EL FUERTE [to say that she had seen] the light descend and lose itself behind the treeline. An hour later, at 22:30, two other witnesses saw the phenomenon on Route 20, near the Aeroclub. The object moved slowly westward. This sighting can be added to others, which have been taking place in our area in recent weeks

Arguably, the defining event of the flap took place in early July:

SOURCE: El Diario del Sur de Cordoba-Villa Maria
DATE: Tuesday, July 9, 2002

Cows Found Inside a Water Tank The Strange Case of Suco, Investigated by El Diario

On the last Friday of June 2002, an event of truly strange characteristics took place in a field of the locality of Suco, located to the west of Rio Cuarto, very near the border with San Luis Province. In the aforementioned Cordoban locality, a well-known livestock producer respected by all for his responsibility and honesty, found 19 dead animals within an Australian-type water tank (translators note: steel-sided, sheet metal tank with a conical cap). Nine of the bovines were dead, according to subsequent medical-veterinarian examinations due to asphyxiation through immersion. The rest were alive, but affected by the low temperatures and near dead due to freezing.

The news not only spread like wildfire throughout the area: it was confirmed by police officials of Regional Unit 7, headquartered at Rio Cuarto, who took over the investigation of the case employing personnel from the Sampacho District Sheriff's Office, located 50 kilometers west of Rio Cuarto on National Hwy. 8.

What no one could explain is how the 19 animals could have entered the enormous water tank, bearing in mind that they first had to cross an electric drover (sic), then a 1.50 meter tall fence, and finally, "jump" over the tank wall.

Two days alter, in a field bordering the first one, the farmhands and owners of the property found a cow that showed the same signs of mutilations suffered by bovines in ranches of Buenos Aires, La Pampa, Mendoza, Southern Cordoba, etc. This time, the mutilated animal had given birth to a calf, although it was stressed that only the mother was affected in this case, showing burns and precise incisions in different parts of the carcass, as though experts had deftly used a special type of scalpel. These cases, according to the statements of a veterinarian named Cumin, who lives in Sampacho, have been investigated from the onset by specialists from the School of Veterinary Medicine of the National University of Rio Cuarto, although they have not received any report that allows them to explain what has really happened.

Those who have dedicated themselves to the possible existence of other forms of life, of UFOs and their consequences, posited the challenging possibility that what they term "teleportations" occurred in this case, thus "explaining" how a lot of 19 animals of large size and weight could have been conveyed by an unknown force from a common and accustomed place (a cow pen) to a strange one (the interior of a water tank), an action that is illogical in both method and objective.

If so, these repeated events do away with the efforts aimed at explaining the events and which lay the blame on both "red-muzzled mice" and "carnivorous bees" and "cattle rustlers". It was acknowledged yesterday that a pasture owner in La Silleta, Province of Salta, found a dead pig showing the characteristics of a mutilated animal, lacking maxillaries, tongue or eyes.

This is how veterinarian Juan Carlos Gimenez Monje discussed the subject. He visited the farm after receiving a call from the owner and added: "I took the animal over to SENASA in Salta and they sent it off to Buenos Aires." The veterinarian explains the situation as being due to the fact that "the agricultural producers stopped using herbicides because of the high cost, then vermin and wild animals can reoccupy the fields once more. Agricultural activity," he added, "consists of zero farming nowadays, and the plough, which used to destroy rodent burrows, no longer enters the fields."

What the expert could not explain was how wild animals could extract the organs in the fashion detailed earlier, with precise cuts, and further having cauterized the wounds.

Following the incident of the water tank which was the first of the Argentine episodes to be reported in the mainstream press across the world, the incidents came in thick and fast. Whereas the first animals to be killed were predominantly ungulates, now, it seemed that any species was able to fall victim to the mystery predator:

SOURCE: Diario "El Ancasti" (Catamarca, Argentina)
DATE: July 10, 2002 MUTILATED DOG FOUND IN CATAMARCA
HIGH STRANGENESS:

It was found in a backyard.

A large mixed-breed dog was found dead yesterday morning in by its owner in their dwelling's enclosed back yard. The animal is missing its tongue, eyes, trachea, aorta, part of its extremities and most of its skin, which appears to have been sheared off by a sharp object rather than torn. There were no signs of blood in the area.

In *The Rising of the Moon* we were very scathing about a notorious series of reports in *UFO Magazine* by researcher Tony Dodd. He presented a series of pictures of mutilated animals. We quoted veteran Fortean Kevin McLure who wrote in 1997:

"Tony Dodd, a former police sergeant turned UFO researcher, has amassed a large number of mutilation reports in Britain. Many of his cases show much the same hallmarks as those reported in the US and other countries - bloodless wounds; the 'surgical removal' of organs, eyes and tongues; the rectum 'cored out' and the jawbone stripped of flesh. Since the early 1990s, Dodd claims, incidents of mutilation have involved wild animals such as foxes, deer, badgers, seals and wild birds, as well as livestock."

Illustrating this piece is a colour photo of a dead fox with a hole in its forehead, and this photo also appears in two very similar articles by Dodd in the July/August 1995 and March/April 1997 issues of UFO Magazine. In the earlier article, the picture is captioned "A fatally wounded fox discovered on a moor at Staintondale, North Yorkshire, in 1993. The hole seen on the top of the head is not a bullet wound. The brain was removed."

In the later article, there are also colour photos of a dead lamb, two dead deer, and a dead hedgehog. The page where they appear is noted "All photographs copyright Quest Picture Library/Anthony Dodd", and the caption reads, "Numerous types of animals both large and small have been mutilated since 1993. This small selection forms just part of a major file currently under investigation by Tony Dodd. The fox, deers, and lamb were discovered in a forest in North Yorkshire between 1993-95. The tiny hedgehog was found in West Yorkshire".

From my examination of the pictures in question it seemed certain that most if not all of the mutilations seen were the result of the predations of invertebrate micropredators such as burying beetles and maggots. Thankfully we were not presented with a parade f similar reports from Argentina. As the reports continued to arrive on our electronic doormat, the predations appeared on the face of it to be uniformly within the descriptions of the Animal Mutilation syndrome as reported for nearly forty years.

The next two reports were typical:

SOURCE: Diario "El Fuerte" (Chascomús, Prov. of Buenos Aires)
DATE: July 11, 2002
BREAKING NEWS

In the Don Cipriano areaAnother Mutilated Cow Discovered

The subject of cattle mutilations appeared to have come to an end following certain professional explanations, but a new case has become known in our district. This one occurred in the "El Taray" field owned by Felipe Sallenave, located on the road to Vado near Public School 19 in the town of Don Cipriano, where the foreman, Fernando Ulloa witnessed to his amazement that a pregnant cow that was in perfect health the day before, showing no signs or symptoms of any malady, was found dead the next morning with signs of mutilation. The specimen was a half-breed Aberdeen Angus and Hereford. It was missing its ears, part of the jaw, tongue, one of its eyes and the nipples on the udders had been shorn off. Part of the rectum was also missing. It was learned that the incisions observed were similar to the ones seen in similar events occurred in other provinces. The pasture field's owner does not consider it possible for the oft-mentioned rodents to have been involved in the act, confirming that he made no police report on the event since he didn't consider it proper (sic). He further explained that access to the field was rather complicated given the poor condition of the roads

SOURCE: Semanario Colón (Internet Edition)
DATE: July 12, 2002
COLON JOINS THE MUTILATED COW CLUB

The bovine, sporting non-traditional injuries, was found in a pasture field located in a field of the Boulevard 17 extension. It was mutilated and this time cattle rustlers were not to blame. The government drafted an official report blaming the shy red-muzzled mouse. By comparison, popular tradition suggests the Chupacabras or little green men.

Whether it is one thing or another, a rancher from Colón had the fright of his life. His name is

Forti, and he lives in a development with 150 dwellings. The pasture field is located between Colón and Wheelwright, and there it was that the owner found a bovine with strange mutilations. The parameters followed by the alleged mice in causing the wounds are identical to the ones repeated in hundreds of cases in Córdoba, Santa Fe and Buenos Aires: precise cuts on the chest, the tongue extracted with admirable precision and the absence of genitalia. The rancher summoned an expert and was further astonished by the find, because nothing similar had ever happened before. This case can be added to the mutilated animals found at Hughes, Wheelwright and Villa Constitución

Then came what the researchers eagerly watching the news for further developments had been expecting for weeks. The first alleged UFO photograph. It came from the neighbouring country of Chile which had been undergoing a wave of strange animal sightings and other apparently disparate Fortean phenomena for several years. As can be seen from the reproduction accompanying the following news item, the image is very blurred and tells us next to nothing:

SOURCE: El "Diario Austral" de La Arucanía - Chile
DATE: July 12, 2002 UFO CAUSES SENSATION THROUGHOUT VILLARRICA

Around 21:00 hours on Wednesday night, dozens of families in Villarrica were watching TV or listening to the friendly voice of their local radio station, which mixed with the soft sound of the rain falling in the darkness and silence of the evening. But this tranquillity was suddenly and unexpectedly altered by the appearance of an unidentified flying object (UFO) which lit the winter night. In spite of the fact that the clouds obstructed a clear view of the object, it was seen from several points of the lake region by hundreds of people, and yesterday, the incredible phenomenon was being discussed in all corners of this city. On that night, in the control booth of Radio Apumanque, Patricio Castillo, 28, was choosing music selections for hundreds of listeners while the rain continued outside. When the clock marked 21:15 hours, the station's phone began to ring.

the radio controller raised a headphone and heard the voice of his co-worker Pilar Castillo, asking him to look out the window because there was a very strange object in the sky. Patricio hung up and opened one of the windows, leaned out and saw the strange object in the dark and rainy winter night. "I looked and I saw a luminous object with three red lights in one of its sides. It was still for a few seconds and then made a very swift movement. I stopped looking at it for a few seconds because I ran in to play a song and when I returned it was gone," he explained. He had been working for 9 years in radio and had never had a similar experience. "I used to not believe in these things, but my opinion changed after this." For one moment, Patricio thought it could be the millionaire who is crossing the world by balloon, but then realized that the had already passed over South America, adding that despite its luminosity, the UFO made no noise, dismissing the possibility that it could be an airplane. Pilar Castillo, the host of the "Morning Talk with Pili" segment, stated that after 21:00 hours she had left a meeting when she realized that there was something strange in the sky. Some of the persons with her were astonished as the luminous object vanished suddenly. "It was like a shooting star that changed colors as it went away. It came from the direction of Nancul to Villarica. It was red for some minutes and then turned violet.

"We left the station and under a southern rain headed to the house of Alberto Sandoval, 47, a self-taught cameraman with years of experience, who managed to record the UFO. Upon reaching the property, one of his neighbors told us we could locate him at Radio Pianisima. We reached the station and found him hosting the "Mundialmente Mexico" program, which airs for one hour every morning.Between Mexican ranchero songs, Alberto Sandoval, who has spent some 28 years in radio, confirmed that he had indeed managed to film the UFO. "A neighbor called me and told me go outside with the camera to film a strange object in the sky. I took out my machine and between the raindrops managed to obtain some images which have a duration of some 5 seconds." This isn't the first time, said Sandoval, that he has obtained UFO images in the lake region. "In the summer of '95 a neighbor woke me up at 5 a.m.. I took my camera and filmed a strange object. The good thing is that the sky was clear. I could see a luminous sphere similar to last night's, only larger."

Within 24 hours UFO fever had reached Argentina with another series of sightings:

SOURCE: Paralelo 32 - Digital Edition (Crespo, Entre Rios, Arg.)
DATE: July 13, 2002 UFOs on the Prowl in Entre Rios

In the past two weeks, "strange lights' were seen in various locales in the center of Entre Rios Province. Residents of Sola, Rosario del Tala and Mansilla were witnesses to the low-altitude maneuvers of objects they were unable to identify with known phenomena such as airplanes, meteorological events, animals or other elements. Last Saturday night, a resident of Parana filmed 15 lights which spun around in circles for 10 minutes. His home video was seen by some news media as proof of the strange phenomena. The strangest case involved police officers in two squad cars at the truck stop on the crossroads of routes 39 and 6. The officers, according to a description offered by Diario Uno de Parana, saw how a powerful light made maneuvers in the night skies on Monday, July 1st. It approached the squad cars, producing "sparks similar to those of a photographer's flash" - a colorful spectacle full of "admirable" color bursts. At a given moment, the light pulled back and the vehicles stopped operating. The engines wouldn't turn over and the lights were off. Half an hour later, when the strange luminosity had vanished completely, the police siren turned on "all of a sudden" and the engines responded to the ignition. The five occupants of both Renault 19 vehicles, one from Rosario del Tala and another from the Gobernador Sola sheriff's office, could find no possible explanation to what they saw. When the object came excessively close to the vehicles, the law enforcement personnel reached for their sidearms, but it seems that the strange light was not impressed, since it continued to fly close to the squad cars pursuing it, withdrawing in its own good time, after the photographic flashes.

Despite the sightings of unidentified flying objects, or perhaps because of them, the killings continued:

SOURCE: Salliqueló On Line (Breaking News)
DATE: July 14, 2002

NEW ANIMAL MUTILATIONS IN SALLIQUELO
**Mouse nowhere to be seen

**A calf was found with incisions identical to those in cases which have already become known. What is strange about the find is that the calf, property of Aniceto Fernández, was found dead and mutilated on a farm adjacent to his own. The owner believed that the animal fled to the neighboring property when it became frightened by something. This is the first case reported in which the animal is in an area of easy access, close to population centers and to any possible witnesses.

Meanwhile, the phenomenon continues, with daily reports of new mutilations taking place in various parts of the country.

SOURCE: Diario "El Argentino" (Gualeguaychú-Entre Ríos)
DATE: Sunday, July 14, 2002 DEAD SHEEP FOUND WITH HEAD MUTILATIONS AT SALTO (URUGUAY)

A resident of Salto is still astonished after finding mutilated sheep

Two dead sheep were found dead on the property of Walter Antonio Remedi, located at Colonia Gestido. The man found a dead sheep with mutilations on its head as he surveyed his fields. The animal found by Remedi was missing an ear, an eye, the tongue and all of the flesh on the left side of the head. A swath of black fur and a lump of fat were found at its side. 500 meters away, a second dead sheep was found with similar injuries to the head area, but it had already been attacked by dogs or other animals. As regards the other animals which have turned up dead in other parts of the country, this case involves sheep and not bovines. There were no signs of spilled blood.

SOURCE: Diario "El Argentino" (Gualeguaychú-Entre Ríos)

By this stage two important things had happened. The Argentine Government's attempts to squash the stories had failed, but they had succeeded in deflecting blame away from themselves. The story above is just one of many to openly mock the theory regarding the mice, and if one is a conspiracy theorist who believes that the whole affair was either organized by, or at the least covertly acceded to by the Argentine Government, it would be tempting to theorise that they had "set SENSA" up by persuading them to come out with such a ridiculous theory when it was obvious that they would meet with nothing but ridicule.

However, I would not deign to suggest that this was the case.

The next step in the saga attracted newspaper headlines all over the world, but did no more than bring knowing smiles to the visages of Fortean researchers worldwide who were expecting something of the sort.

Whenever there has been a major wave of animal mutilations (or indeed any other quasi-fortean occurrences) there have been sightings of strange animals which defy rational scientific categorization. In *The Rising of the Moon* we present accounts of a large bioluminescent felid (a zoological impossibility) alongside animal mutilations and UFO reports in Exmouth during 1997. In *Only fools and goatsuckers* I explain the genesis and evolution of the imagery of the chupacabra, and touch on the subject further in my forthcoming book *Monster Hunter*.

No-one at the CFZ therefore was particularly surprised when the first reports came in of a strange being, reported in the midst of the mutilation episodes:

SOURCE: "El Diario" de La Pampa
DATE: 16 July 2002

WOMAN SEES GREEN DWARF AND FAINTS

A young woman fainted yesterday from nervous shock after having an alleged close encounter with "some sort of green dwarf" that appeared before her on 23:30 hours two nights ago as she walked - enveloped in darkness - along Misioneros Salesianos street only meters away from Raul B. Diaz St. in the Villa Elisa neighborhood. This information was provided to El Diario by police sources in the Second District, stating in regard to the incident that "the protagonist's identity, and that of the other three women who attended her until the ambulance arrived, is not in our hands. We do know that at 11: 30 at night a request for help was received from the corner of Raul B. Diaz and Salesianos and a squad car headed toward that location."

According to the source, it was learned that "when personnel reached [the scene], the women were not there, in any event, the duty officer established that one of these persons saw some kind of green dwarf that looked at her pointedly, and she fainted."

Subsequently, a reporter for this paper was able to find out, from medical sources, that the woman was treated by personnel from an EMS ambulance under the direction of Dr. Brum, who "revived the protagonist and helped her out of her state of nervous shock. She was later sent home and the ambulance returned to the hospital where the doctor and paramedics were on duty."

Almost immediately the newspapers drew the inference that the two sets of phenomena were somehow linked:

STRANGE DWARF REPORTED IN CATTLE MUTE AREA GENERAL ACHA (DNA) –
The presence of a "dwarf" or "green midget" wandering the backyards of many neighborhoods in General Acha has been the latest subject of conversation. It appears to be a short, green entity which runs away with haste when detected.

A resident of the El Oeste neighborhood informed this newspaper that the entity appeared twice on her property: once in the morning and once in the afternoon. According to the woman, her husband was resting on both occasions, but when he reported to the backyard in response to his wife's screams, the mysterious entity vanished by rapidly climbing up a tree. While the man did not see the "dwarf", he gives his spouse the benefit of the doubt, stating that he has known her for 40 years and can vouch for her physical and mental health. All versions agree in that the entity is short in stature. Another resident added that it moved so quickly that it was hard to describe it. The fact is that this entity, described as "a green dwarf", is causing a stir in the city and its manifestations have been the subject of conversation in certain circles. The stories have already captivated radio audiences and interest and curiosity for the subject increase as time goes by.

It is perhaps surprising that nobody seemed that surprised that the triune of phenomena: The UFOs, the dwarf and the animal mutilations were now spread out over three countries. Although the dwarf itself had been seen in the animal mutilation heartland of La Pampa, reports cited above had been from as far afield as Uruguay and Chile.

Still the killings continued:

SOURCE: La Voz de Bragado (year 2, No. 793)
DATE: Wednesday, July 17, 2002

A Mutilated Calf found in Bragado

Found in a field located in the town of Rauch Viejo Died a natural death and was found mutilated 2 days alter

A calf was found mutilated in the town of Bragado (Prov. of Buenos Aires) and the subject was discussed by several local residents. The event was confirmed yesterday morning and it took place in a ranch located in the Rauch Viejo quarter, more precisely at the Estancia Maria Magdalena, owned by the Vaccarezza family.

Our newspaper reported to the scene and in a conversation with Mr. Oscar Latorre, we were led to the site where the lifeless animal lay, and where it was found by Latorre himself.. He explained that the animal died approximately 15 days ago, apparently intoxicated. "The animal died constipated," he remarked. He stated that he was able to see the animal with signs of mutilation on Monday morning and that he has no doubt "that it could be a rodent, given the tracks I found."

What is most extraordinary about this case is the fact that the animal has been dead for 15 days yet "does not present any signs of decomposition and has not been devoured by vultures nor foxes, who are frequently seen around this area, " said Latorre. On site, we were able to ascertain that the animal was missing its tongue, eyes or ears (Photo 3) and had perforations in the anal area.

By this stage in the Flap, events were taking a familiar turn. Despite the horrific nature of the mutilations on the dead animals, both journalists and researchers alike were seeming to get immune to the cavalcade of horror which was being paraded before them almost upon a daily basis. The news report immediately above is a case in point. The descriptions off "perforations in the anal area", with the gamut of sexual, social, and biological implications that the sentence gives, is delivered in such a matter-of-fact manner that it is easy to overlook. Despite the fact that once again the SENSA Report explanation is ridiculed, by this time even SENSA had ceased to be the enemy, and the journalist a limited himself to a brief and matter-of-fact description of what had happened. Something particularly interesting about this case, is that the cadaver is described as not having become putrefied. This apparent incorruptibility and is something which many animal mutilation cases describe. It is something, which at the moment defies any rational explanation, although it has been, suggested that if these animals were killed as part of some biological experiment then they could have been injected with some re-agent or preservative. To the best of my knowledge no such chemical has ever been found in any victim.

The subsequent spate of killings continued to be reported in a similarly matter-of-fact manner:

Jul 23, 2002

SALIQUELLO, Argentina - Daniel Belot has seen his share of dead cows.

As a veterinarian in the heart of the cow-full pampas, Belot has written off bovine deaths to causes as diverse as foot-and-mouth disease, bloat, lightning, killer bees and cattle thieves who butcher their loot in place, a crime that has become increasingly common as Argentina's economic crisis has extended to the countryside.

Then, in April, he discovered a case that stumped him. A rancher had found a nearly 1,000-pound Aberdeen Angus lying on its belly "like a rabbit," in Belot's words. The left side of its face around the jaw was gone, the hide cut away in two straight lines meeting at a 90-degree angle.

Its tongue, pharynx and larynx were missing. Muscles and ligaments had been removed from the jawbones, leaving them spotless. There was no blood on the animal or nearby; nor were there signs of scavengers or predators.

I had never seen anything like it before," says Belot, who works for Argentina's animal health agency, Senasa, in the sleepy town of Saliquello. "How were those cuts made? When? Why?"

Three months later, Belot has no answers. Across this country's immense, grassy plains, Argentina's renowned beef cattle are turning up dead, mutilated in ways that have baffled experts and spooked ranchers.

Since Belot detected the first mutilation in April, nearly 200 more have been reported in the area, in addition to a scattering of cases from as far away as Patagonia and Uruguay.

Most cows have the same missing parts as the first one examined by Belot. But all the mutilations share an uncanny similarity: Organs, flesh and skin have been removed in angular or neatly curved cuts that leave no blood and clean, dry bones.

"he type of incisions do not coincide with any infectious or contagious disease that we know," says Alberto Pariani, a veterinarian at the University of La Pampa who has examined 40 mutilated cows. "When animals eat, they rip, they tear. They don't cut."

Everyone who has experience working on the ranch says the same thing: No animal can do this."

Blame has been pinned on everything from ravenous rodents to satanic cults, but in the farmhouses and small towns that dot the pampas, the paranormal is the No. 1 suspect. Sure enough, the mutilations have been accompanied by a spate of UFO sightings.

The mutilations are not without precedent. Since the 1960s, hundreds of mutilated animals have been found in the United States with nearly identical characteristics - removal of organs in what appear to be surgically precise cuts, no trace of blood, no tracks of humans or animals, often with coinciding testimonies of strange lights.

Mutilation cases have been reported during the past year in Montana and Oregon. The news media have evoked comparisons with the legend of the chupacabras, literally "goat sucker," revived in Puerto Rico several years ago when farm animals there were reportedly found dead and bloodless with abnormal puncture wounds.

But according to an Argentine government-backed investigation, the mutilations have an earthly explanation. A team of university veterinarians working with specialists from Senasa and the National Institute of Agricultural Technology recently announced that they had caught the mysterious cow mutilator.

The culprit's name: Oxymycterus, commonly known as the long-nosed mouse. The theory holds that the cows die from disease or other natural causes, not unusual in winter, and are then set upon by scavengers, including foxes and birds. But it is the hungry long-nosed mouse, with its four potent incisors, that is allegedly responsible for nibbling off flesh and hide in circular and linear cuts.

To prove their hypothesis, veterinarians at the national university in the city of Tandil placed dead cows in areas where some of the mutilations had been discovered. Four or five days later, the cows were left with "lesions exactly the same" as those discovered in the mutilated cows.

The announcement was made at a Buenos Aires news conference, where reporters were shown a video of mice crawling through a carcass and chomping a cow tongue on a laboratory table. National media coverage of the mutilations has effectively ended since the news conference.

But many experts and local veterinarians remain unconvinced by the government-endorsed conclusion. One question the university team has not answered is why the mice, or any scavenger for that matter, would consume the hide around the jaw instead of first devouring the rest of the cow's softer flesh and innards.

Another chink in the theory: Some cows have been found mutilated hours after being seen intact, leaving scant time for the mice to remove the organs.

Nobody, from ranchers to biologists specializing in rodents, has ever seen mice feed on a cow carcass.

he Tandil veterinarians suggest that a demographic explosion combined with an unusually cold winter have driven the mice to change their diet from worms and slugs to cow flesh. But in many cases, witnesses have seen no signs of mice or any other scavengers. Raising even greater doubts, the long-nosed mouse does not inhabit the province of La Pampa, where dozens of mutilations have been found.

The team of university and government specialists limited their study to five counties in the province of Buenos Aires. They did not make available the details of their investigation.

But if the mouse theory has its holes, the possibility of human involvement seems even more unlikely. Police have found no footprints or tire tracks near the animals. Nor are there signs of struggle; cows killed by predators or humans usually leave kick marks as they take their final gasps. In some cases, the cows were discovered behind fences and locked gates or miles from the nearest road.

Nobody has seen anyone or anything out of the ordinary, except weird lights in the sky.

"We are totally disoriented," says Oscar Raul Arce, the chief of the provincial police in northern La Pampa. "What is really striking is that no clues, or prints or blood have been left.

"What's going on here is perhaps beyond our ability to understand."

For most people out on the pampas, where cows outnumber humans in the range of 10-to-1, something strange is responsible for the mutilations, and it's not the long-nosed mouse.

"I'd always heard stories of people who had seen lights and strange things," says rancher Raul Vargas, 39, standing over a mutilated calf found the day before a half-mile from his farmhouse.

"But if I hadn't seen this with my own eyes, I wouldn't have believed it."

The SENSA report has once again been wheeled out, but by this time even the most hard-nosed journalists couldn't be bothered to show how ludicrous it was. By this stage at the political and economic situation in Argentina was verging upon anarchy. There were riots in the streets, the financial institutions have either collapsed or were on the verge of doing so, and in a massively controversial move the Argentine government arrested the former dictator General Galtieri, planning a to put him on trial for human rights abuses. This socio-political U-turn, during which they denied one of the major incidents of their recent history, must have been a hugely confusing to the population at large. The man who would once been their national hero, was now condemned as a war criminal. In my book *"Only Fools and Goatsuckers"* (2000), I described the 199/6 wave of chupacabra attacks across Mexico against the background of political unrest in the region, and feel certain that in the same way as the chupacabra incidents were somehow directly related to the groundswell of popular support for Sub Commandante Marcos and his FZLN Guerilla separatists, the socio-political events in, Argentina and the outbreak of animal mutilations are somehow related.

There is only one more report in our files for July:

SOURCE: El Comercial (newspaper)DATE: Sunday, July 28, 2002 BETWEEN BARTOLOME DE LAS CASAS AND FONTANA Five Cows Mutilated

Four mutilated cows were found again some 12 km southwest of Bartolome de las Casas, having the same characteristics as earlier cases. Two were found on Wednesday and two more on Thursday at a field owned by Ariel "Papin" Suarez. The fifth animal was found yesterday in Estanislao del Campo: the victim was a cow belonging to Victor Andres Roldan in the Ranero Cué pasture field, some 10 km. distant from the town, having all of the characteristics indicate, such as "mutilation," "cauterization" and "millimetric incisions." Fourteen "unofficial" reports were received in the Fontana and Bartolome de las Casas region; reports include a young horse, a buffalo and 12 bovines. A more extensive report will appear tomorrow.

If this report did appear, we have no record of it.

At the beginning of August it seems obvious that the authorities felt that they had to make a firm stand. The next report, shows the beleaguered civilian authorities, coming under fire from a population who believe that they are doing nothing about the crisis. The police and the regional government officials began coming forward and making statements on a more regular basis, and it seems logical to hypothesise that they were doing so as a direct reaction to levels of unrest amongst the rural civilian population. The next news item is very much a snapshot of an administration who really did not know what they had to do next:

SOURCE: El Diario de La Pampa DATE: Wednesday, August 7, 2002

CONTRADICTIONS EMERGE IN REPORTING ANIMAL MUTES

"There exist no orders to cease accepting reports on cattle mutilations," said the chief of Regional Unit 1, Sheriff Inspector Juan Carlos Gorris, as well as the Assistant Chief of R.U. IV in the town of 25 de Mayo, Deputy Sheriff Osvaldo Olie, in regard to public statements by rural livestock producer Julia Cavanah, who said: "Local police do not take down reports involving mutilated cows, because there are orders from above [not to do so]."

"Under no circumstances can we stop accepting reports, because we are obligated to act. The judge will determine later on if a crime should be investigated or not," said Gorris.

On this issue, the Chief of R.U.1 indicated that "today (meaning yesterday) I was in touch with the chief of the Telén Dependency (deputy sheriff Hector Bucher) and I asked him to go to the El Nauco ranch, 120 km away from said town, to make the pertinent inquiries."

While adding that the did not have Bucher's report in hand, he pointed out that "no one has gone to the sheriff's office to file a report on these cases."

He then explained that El Nauco ranch is 50 hectares in size, and in this regard, "part of it falls under the jurisdiction of RU1 (at the Telén Sheriff's Office) and the other under RU IV (the Chacharramendi police station). The officer in charge of RU IV, Osvaldo Olie, indicated that "we are in the middle of an investigation. We still do not have official information to give out, because we're just finding out what happened."

Julia Cavanagh is the daughter of the pasture field's administrator, located some 70 km west of Chacharramendi, where she said that since June 20th to date, "six cows and four bulls have been found mutilated."

The next report shows the authorities desperately trying to do something. It is interesting to note that in the previous report the police were unwilling to speculate as to whether a crime had been committed on not. When faced with an epidemic of this size is difficult to know what else they could really have done. Once again SENSA/SENASA are called in and this time, presumably because of the sheer scale of events in question, the newspapers have ceased to ridicule them:

SOURCE: Telam News Agency (Argentina)

DATE: August 9, 2002
MUTILATED BOVINE FOUND IN CATAMARCA

A mutilated bovine, missing its genitals, tongue, ears and eyes was found in recent hours at the El Peñon location in Catamarca near the El Rodeo summer village, 38 km from the provincial seat, according to the National Health and Agro-Alimentary Quality Service (SENASA).

This agency confirmed that a commission is working in the area to reliably determine the appearence of a mutilated bovine. According to the police report, it presents characteristics similar to those of other bovines found in other Argentine provinces.

The discovery occurred at Cerro Ambato, according to the owner of the mutilated cow - Carlos Sosa - who later made the corresponding claim at the El Rodeo deputy sheriff's office for the alleged crime of "damages". Police confirmed that according to Sosa, the animal in question is a cow missing its genitals, eyes, tongue, and ears, and has a hole in the area of the knee, yet shows no signs of decomposing.

Specialized personnel from the National Institute of Livestock Technology (INTA) and SENASA-Catamarca decided to transfer the rest of the mutilated cow to the Catamarcan capital to have them analysed in a lab.

With the three burros found this week in the province of Jujuy, the number of animal species being mutilated rises to seven. Aside from bovines, which are involved in 90 per cent of the events, horses, pigs, sheep, guanacos and wild boar have also been mutilated.

Another report from the same day shows the newspapers and a desperate attempt to find experts who can throw some light upon the this ever more intriguing and revolting mystery:

SOURCE: Diario La Nueva Provincia - DATE: August 9, 2002
THREE BURROS MUTILATED IN JUJUY

JUJUY - The discovery of three dead burros, showing signs of mutilation, raised concerns among residents of El Churcal in the Jujuvian department of Humahuaca, according to police sources.

The animals, belonging to Miguelina Martinez, were found lying a pasture field. One of them was missing an eye, another its tail and part of the anus, while the third one - a pregnant female - had a circular wound in the lower abdomen and was missing its foetus.

The discovery of the mutilated animals was corroborated by Sixto Vazquez Suleta, an author from Humahuaca and former cultural affairs director of the province, who stated that the case is not normal and is not related to normal predator activity.

Vazquez Suleta stated that the animal bodies give off no odor, while dogs refuse to approach them, and there are no signs of violence or struggles in the place they were found. The mutilation claim was made at the Humahuaca sheriff's office, whose headquarters dispatched a mission to scene to learn further details.

August 9th was probably the most busy day in the history of the epidemic. The authorities have reacted to this latest spate of killings by sending in as many investigation teams as they could muster. It is interesting to note that in the previous item it is once again hinted that these corpses are not subject to putrefaction - they give no odour, and for some reason neither dogs or other predators approach them. Still the killings continued:

SOURCE: Telam News Agency (Argentina)
DATE: August 9, 2002 MUTILATED BOVINE FOUND IN CATAMARCA

A mutilated bovine, missing its genitals, tongue, ears and eyes was found in recent hours at the El Peñon location in Catamarca near the El Rodeo summer village, 38 km from the provincial seat, according to the National Health and Agro-Alimentary Quality Service (SENASA). This agency confirmed that a commission is working in the area to reliably determine the appearence of a mutilated bovine. According to the police report, it presents characteristics similar to those of other bovines found in other Argentine provinces. The discovery occurred at Cerro Ambato, according to the owner of the mutilated cow - Carlos Sosa - who later made the corresponding claim at the El Rodeo deputy sheriff's office for the alleged crime of "damages". Police confirmed that according to Sosa, the animal in question is a cow missing its genitals, eyes, tongue, and ears, and has a hole in the area of the knee, yet shows no signs of decomposing. Specialized personnel from the National Institute of Livestock Technology (INTA) and SENASA - Catamarca decided to transfer the rest of the mutilated cow to the Catamarcan capital to have them analysed in a lab. With the three burros found this week in the province of Jujuy, the number of animal species being mutilated rises to seven. Aside from bovines, which are involved in 90 per cent of the events, horses, pigs, sheep, guanacos and wild boar have also been mutilated.

In the midst of the killings, the newspapers and other commentators looked around desperately for somebody or something that they could blame as a culprit. The next report, from Bahia Blanca, an area well known to forteana, may or may not be true. It is most important and not because of what it says but because of when it says it. The civilian authorities, the police, the military and SNSA/SENASA had not managed to find a culprit larger than a mouse. May be, instead of looking around the Earth for someone to blame they should start watching the skies:

SOURCE: Diario "La Nueva Provincia" (Bahia Blanca, Argentina
DATE: August 9, 2002

RANCHER CLAIMS BEING PARALYZED BY UNKNOWN OBJECT

**Reports on strange cases emerge

**The events involving mutilated animals, which re-emerged in recent days, and the UFO sightings at Punta Alta, are now joined by the strange story by a rancher from [the town] of Jacinto Arauz, who claims having been only a meter and a half away from a "flat, three-legged apparatus" - a contact which left him immobilized for an undetermined span of time.

Claims of strange lights crossing the skies at night and even the account of a rancher who claims having had a close encounter in broad daylight have joined the mutilations in this locality, which began in early June [2002]. This is the case of Raul Dorado, owner of a field near this location, who was the protagonist of the startling event, which took place as he toured his field, which he does on a daily basis.

According to his story, everything began with a noise similar to that of a "whirlwind" and he later saw how a strange craft descended over him, whisking away his cellphone and leaving him momentarily speechless. The rancher was born and lived in the countryside until up to ten years ago.

Today, a resident of this Pampan community, he visits the place every day. The site covers only a few hectares located 4 km north of the city, on the sides of Route 35, with groves of trees and reeds growing at the edge of a stream crossing the pasture field. "As I was heading back toward the farmhouse, where I'd left the car, was when that thing appeared, " he recalls.

"It was a dim green colored circumference - although in daylight, I wasn't able to note its luminosity. It had three legs and was barely a meter or meter and a half away from me," he maintained.

"When I saw it," he continued, " I felt something like an electrical charge. I fell to the ground on my knees, paralyzed, unable to do anything, with my shotgun leaning on the ground. I was also carrying binoculars and had a cell phone in my left hand, which was sucked up by the green circle. I saw it rising and disappearing." The following is a segment of the interview between "La Nueva Provincia's" reporter and Raul Dorado. - What was the object like? - -Given that I could only see the bottom part, it was a smooth surface with three legs. At no time did it make contact with the ground. I don't know how many minutes it was there, because I lost all notion of time. - -Did you see any figures or persons descending from the craft? - No, all I saw was my cell phone "traveling" toward the craft. The same did not happen to my weapon or my binoculars. - How did the object vanish? - It took off toward the west with a muted sound, as though it from a strong wind, a whirlwind that occurred instantly and at high speed. - How did you get back into town? - It was roughly 17:00 hrs. when the event occurred, and it took a lot [of effort] to reach the farmhouse and the car. I left, and came back home at more or less 19:00 hrs.. I tried telling my wife but I couldn't speak.....

"I always tell him to come back by daylight," added his wife, Elda, who stated that she was starting to get concerned since her husband wasn't home yet." I was about to dial the cell phone when he turned up. He tried to speak but couldn't, and he hugged me and started to cry. I called my neighbor to take him to the hospital, and he then he scribbled what happened to him on a piece of paper, which was later taken by the police."

He was attended by physician Ana Maria Lazaric at the hospital, whose examination found nothing abnormal except for a puncture mark in the left ring finger, between the nail and the flesh of the cuticle. How it occurred could not be determined. "He was in a severe state o shock and the doctor gave him a sedative. He wanted to come home and as soon as he woke up the next morning he was able to speak," explained his wife.

Dorado said that the puncture was not on his finger prior to the event, but was unable to explain when the injury occurred. "The ring and middle fingers of the hand in which I held the cell phone still hurt. My jaws, shoulder and legs also hurt. I always wanted to see something like that, but nothing like this had ever happened to me."

While no mutilated animals were found on his property, he explained that a mutilated cow was found 200 meters from his field last week. "A few meters away from the site there's another [mutilated cow] and some 300 meters southeast they recently found one a few days ago. Now I'm more convinced than ever that these mutilations are not caused by mice or other animals."

Senior officer Marcelo Alarcon - in charge of the Jacinto Arauz sheriff's office due to Deputy Sheriff Hector Rosane's furlough - reported that the events at Dorado's field are still being researched. He also noted that the has toured the location in search of tracks or the missing cell phone, although without any results.

Regarding the cell phone, police researchers placed phone calls to it on the same night the incident took place to see if anyone answered. On two tries, it would ring and then cut off, and on a third attempt, a sort of breathing could be heard, and then the sound of the keypad being dialed. Follow-up efforts - constant until the following evening - aimed at finding out who was answering the phone indicated that the cell phone is either off or beyond the coverage area.

Reading and re-reading the above account, it is difficult not to see these other anomalous phenomena as mere adjuncts to the main wave of animal mutilations. This is the first, and as far a we are aware, the only incident during the wave of animal mutilations where there has been any sort of an attack mentioned on a human being. The attack is relatively minor compared to those, which had been reported on the domestic livestock, but nevertheless is a significant development. The hints that some sort strange electromagnetic phenomenon had interfered with the cell phone are, we believe, significant, not necessarily because we believe that they did actually happen. What is more important, to our mind at least, is that these are all events which have been described as happening in during animal mutilation episode waves in the past.

On the same day there were 10 more killings:

SOURCE: Diario La Nueva Provincia
DATE: August 9, 2002
TEN REPORTS IN PATAGONES

PATAGONES (A) - Despite the fact that the red-muzzled mouse was officially fingered by SENASA as the culprit in over 200 animal mutilations in our country, and that the case was closed by said nationwide agency, cases are still being reported in Patagones.

This is what rancher Hugo Valdez told "La Nueva Provincia". Valdez, whose property is 50 km northeast of the district seat and some 5000 meters distant from the first cases, at the Miguel Angel Cordoba field. "My son Ignacio made the discovery while out hunting with a friend. They approached the dead animal when they saw it." According to Valdez, the animal's death would have occurred some 10 days ago, approximately.

Upon reaching the site, they saw that it presented the characteristics of the mutilated animals. Faced with this find, they marked the area, returned home and then went back to see if it was really similar to other cases. "We proceeded to photograph it and perform a necropsy to see what parts were missing," said Valdez.

"We could readily see that in part of the lower jawbone, the bone was entirely desiccated. Right there, in that part, it gave the impression of an animal that had been dead for a very long time, " he added. This was not the case, however, with the rest of the body parts, meaning to say that there was no rigor mortis. On the other hand, the udder had been cut off in a circular manner and the anus had been extracted altogether. The digestive and reproductive tracts, as well as the bladder, were absent. There were no incisions nor extractions to be seen in the abdomen, on the other hand.

"It was also missing its tongue, but not its trachea. We were surprised when we cut the shoulder blade and turned [the animal] around. There, the flesh seemed practically natural, since an artery was cut and the blood was coagulated. We then opened the abdomen and the rumen, and we could see that the dung was as one would find in a recently butchered animal," explained the rancher.

He noted that there was a certain degree of odor, but the flesh gave the impression of having all of its colors, as though the animal had been recently butchered, which can be seen in the photographs

[NOTE: NO PHOTOGRAPHS PRESENTED WITH THE ARTICLE].

Hugo Valdez told this paper that he had comments on other cases, but never had direct access to any animal. "This is the first one on my property, and we don't discard the possibility that there could be others, since the wilderness is very dense and access is difficult."

The necropsy was carried out by Valdez and his son, who is a veterinary medicine student and whose school - in the province of La Pampa - had obtained access to parts from animals mutilated in this province. "What little he could see resembles greatly what we found in the field. Furthermore, a classmate from the Pampan community of Cuchillo Có had several mutilations on his own property and brought samples to the university. That's what they were able to see," concluded the rancher.

Another tank-emptying case occurred in a pasture field near to Hugo Valdez's, according to the property's owner, who refused to be identified. One of the large-capacity "Australian tanks" was found completely empty. The event occurred only a few days ago.

If this come wasn't enough, there were other killings in different parts of the country, all following the same basic template:

SOURCE: Diario La Nueva Provincia - DATE: August 9, 2002
THREE BURROS MUTILATED IN JUJUY JUJUY

The discovery of three dead burros, showing signs of mutilation, raised concerns among residents of El Churcal in the Jujuvian department of Humahuaca, according to police sources.

The animals, belonging to Miguelina Martinez, were found lying a pasture field. One of them was missing an eye, another its tail and part of the anus, while the third one - a pregnant female - had a circular wound in the lower abdomen and was missing its foetus. The discovery of the mutilated animals was corroborated by Sixto Vazquez Suleta, an author from Humahuaca and former cultural affairs director of the province, who stated that the case is not normal and is not related to normal predator activity. Vazquez Suleta stated that the animal bodies give off no odor, while dogs refuse to approach them, and there are no signs of violence or struggles in the place they were found. The mutilation claim was made at the Humahuaca sheriff's office, whose headquarters dispatched a mission to scene to learn further details

As the killings continued, Scott Corales wrote the following letter to an Argentine newspaper in which he pointed out the links between the animal mutilation episodes and other paranormal phenomena which had been reported across the country:

Subj: **Argentina: The Strange "Shroud" of Ciudad Atlantida**
Date: 8/9/2002

Dear Readers of Inexplicata,

The paranormal aspects of the unprecedented UFO/cattle mutilation wave being experienced in Argentina since May 2002 are becoming more pronounced. A few days ago, we posted a newspaper article regarding the strange appearance of a ghostly entity near Ciudad Atlántida. We have the pleasure of presenting you with a follow-up on that story, courtesy of researcher Oscar Adolfo ("Quique") Mario.

Scott Corrales - Institute of Hispanic Ufology

Regarding the news item sent out by Gloria [Coluchi] a few days ago, I would like to mention that the strange figure which many residents now call "The Shroud" is still appearing in Punta Alta, some 27 km from Bahia Blanca. This phenomenon has been verified since Monday, July 29 in an area filled with small sandbanks along the road leading to the Arroyo Pareja municipal beach, facing the Ciudad Atlantida neighborhood, which borders the wire fencing of the Puerto Belgrano Naval Base.

The apparitions have been witnessed not only by neighbors, but also by officers of the provincial police and military men on duty at the Belgrano Naval Base, located some 200 meters from the sand dunes where these creatures are seen. The police has been summoned by these same neighbors to verify the accuracy of their statements. In response to these events, and while they have not wanted to confirm them officially, the authorities have stepped up patrols in the area, since the facilities of the Naval Officers Club happen to be in vicinity.

The apparitions, which always occur in the early morning hours, have become so common that in recent days several neighbors have stayed up late to see them, and it is thus that they were able to confirm the simultaneous presence of up to 2 such beings, as well as red lights flying over the area. All of the witnesses, whether individuals or in a group, agree that the entities are nebulous ("as if made of tulle") and glowing, with a pair of red eyes being clearly identifiable, and glassy faces ("as if transparent").

Their movements are smooth and they always appear from behind the sand dunes, as though "coming from the [Naval] Base". Most witnesses agree that they show a great interest in the water tanks located above the roofs, and that they do not flee upon detecting the presence of local residents - rather, they stare at them fixedly for some minutes. Deep silence is perceived during the observations and "the air appears to become still" and neither heat nor cold can be felt - only a sensation of warmth, as if [the beings] emanated heat. Witnesses also state that things appear to be darker when "the shroud" appears, as though ambient lighting were dimmed. Another curious fact is the large number of cats and dogs who report to the place where the sightings occur and sit down to observe it in silence, in a state resembling a trance.

A series of semicircular prints have been found in the dunes, not very deep, with 15-20 cm separations in some cases and 30-50 cm in others. The locals have identified them as the strange beings' footprints. Efforts have been made to photograph the "shroud", but all images, despite the use of different cameras, appear exposed.

One witness who approached the site to witness the phenomenon took two photos of the prints in the afternoon and decided to stay with the group of people hoping to see the "Shroud", managing to take some photos of same and record the new prints which appeared in the dunes after the sighting. Curiously, while the first images of the tracks as well as the ones taken subsequently were perfectly clear, the intermediate photos in which the strange being ought to have been seen were exposed, which could be seen on the negatives as well. An engineer who visited the site suggested that the apparition perhaps generates some sort of radiation that causes film to become exposed, relating it to the "heat" felt by witnesses when the figure approaches.

A completely uncommon print was found at the site - but the residents aren't sure if it's related to the phenomenon: it has the shape of a five-fingered claw.

One witness tried to make use of his carbine's telescopic sight to see the figure in greater detail, but "when he tried to focus on it, he could see nothing [..] it was as though it could only be seen with the unaided eye."

Other residents suggest that the interest shown in the water tanks may bear some kind of relationship with the disappearance of large quantities of liquid in the "Australian tanks" of Bajo Hondo (some 40 km from the site), which occurred in concatenation with the appearance of mutilated bovines, although there is nothing to sustain this theory.

In our opinion, it is necessary to view the phenomenon holistically. Were as the main physical manifestations are without doubt, the mutilated animals, the UFO sightings, the strange shroud, and possibly most important of all the manifestations of the green dwarf are, we believe, essential, in trying to reach an understanding of the phenomenon as a whole. It would be utterly simplistic to follow the conventional "X Files" explanation that the green dwarf, and the chupacabra before it, were passengers in the intergalactic UFO, and for some reason a which only makes sense if you happen to be an alien from another planet, they have been conducting a nasty experiments on hapless livestock across Argentina. This is, at least as far as this author is concerned, a farrago of complete nonsense.

We believe, but that's when examining the phenomenon of this strange dwarf, one should have a look at the aetiology of the reports of strange beasts in Puerto Rico. As Scott Corales has shown in several of his remarkable books, and as we discovered during our expedition to the island in 1998, the chupacabra is only the latest in a long line of Mysterious were Zooform entities to have been reported from the island. In the 1950s there were bigfoot type animals reported, in the 1970s the vampire of Moca, in the early 1980s a weird porcupine like creature, and finally in the mid-1990s el chupacabra himself. It would seem very likely that this green dwarf is just the latest in a similar line of creatures.

No less a personage than Heuvelmans noted spates of animal mutilation activity in Argentina during the middle of the last century. At the time these were blamed on *didi* or ape-men, and Heuvelmans correlated these reports with other reports of anthropoid apes seen across South America. He drew a link with the famous photograph of *Ameranthropoides loysii* which he hypothesise was a genuine form of South American ape. Recent research by Loren Coleman has, however, revealed these photographs to be a hoax, and furthermore one with unpleasant racist connotations and motivation.

There are a plethora of reports of small hairy dwarfs, surprisingly similar to the agogwe of East Africa stop these were reported at all across South and Central America in the 1960s and 1970s. The chupacabra was the demon of choice during the 1990s and it seems very likely that the new sightings of a green and malevolent dwarf of the Pampas will become more and more frequent as this it becomes the "Fear Icon" for the first decade of the 21st century.

It seems that just as the Argentine economy and socio-political situation began to recover, so the animal mutilation reports died off. August 9th was the last significant day of the episode. The important thing as far as we can discern is not merely to keep a record of each and every dead cow found with unexplained injuries. It is only by seeing the phenomenon as a hall, and furthermore as part of a global continuum which has been going on for at least 200 years, that we can hope to make sense of these distressing episodes.

British researcher Kevin McClure has been very scathing of animal mutilation investigations, and has stated publicly that he does not believe that they are a worthwhile field for Fortean Research. When one realises that he was on referring specifically to the work of Tony Dodd et al it is hard not to agree with him. However when one examines episodes of phenomena such as those described in this paper, this researcher at least, is of the opinion of that these episodes can and must be studied.

They seem to be an integral part of the human condition as the 20th century drew to a close and the 21st century has made a violent beginning. Somehow, these episodes are inextricably linked to the socio and political change in Argentina. Similarly we have shown a link between similar episodes and social, political, cultural, and religious change in Puerto Rico, Mexico, Ireland, the United States, and the United Kingdom. As the global infrastructure becomes more and more unstable as market forces led companies, abhorrent dictatorships, and global catastrophes threaten to overwhelm our poor little world it seems likely that if there is a direct correlation between episodes of the type described in this paper and the sociopolitical change then attacks on domestic livestock again become more and more frequent.

It must be remembered, that the one most likely explanation for these semi ritualised killings has not been addressed in this paper.

Using Occam's razor it does seem most likely that these killings have been done by people. But what is the motivation? It is unfortunate fact of life that some people enjoy hurting animals and indeed other people, but bear in mind the sheer scale of for the killings in Argentina, and the geographical area over will which are these incidents have taken place - over three entire countries, it does seem unlikely that the same people (if indeed it is people), are responsible. What about copycat killings? Again the same answer. What is the motivation?

It is hard to imagine that no matter how disenchanted you are with your government that you would feel that are the only effective way of registering a political protest is to cut up somebody's cattle. It just doesn't make any logical sense. This researcher it would like to dismiss the alien hypothesis out of hand. There is no evidence whatsoever that it we have ever been visited a buyer sentient beings from another planet, and even if we have we are left to the same question - what is the motivation? Like so much in the fortean universe it doesn't seem to make any sense.

If the theories propounded by the present author, and Nigel Wright in for *"The Rising of the Moon* (Domra, 1999), are indeed correct, then it is only by reference to them that a workable scenario to explain the horrific wave of killings all across Argentina during the summer of 2002 can be produced. Either that or SENSA/SENASA were right all along and the whole thing can be blamed on those damned field mice.

The final word on these episodes can be found in the following newspaper report from the

Guardian, which show without any doubt, if any doubt were needed y a researchers worldwide have a such little faith in the British press:
The Guardian 10 Aug 2002

Goatsucker claims dole

The "goatsucker", a mythical figure blamed for mysterious animal maulings in Argentina, has shown up on an unemployment benefit list, entitling it to about £27 of dole money a month. Officials in Catamarca blamed the error on a technical hitch, but said the goatsucker never showed up to collect its cheques.

Reuters, Buenos Aires

In view of comments like this is it any wonder that those of us who write for the popular press quite often do not like to admit the fact.

Kent's exotic cats caught on film

by Neil Arnold

Evidence across Britain for the existence of exotic species of felines is substantial; no other animal on this island kills with such a devastating throat bite, or drags its prey into a tree, or, in many cases of encounters, can be mistaken for something four-feet long with a tail three-quarters the length of the body.

The British exotic cat phenomenon is, although often inaccurately, a well publicised situation which tends to put the welfare of the animals to the background so that those who 'research' them can search for what they believe is a treasure at the end of their personal rainbow. Unfortunately, the felines that roam this country are advertised in the press to the extent that they become hounded, injured and in some instances killed, and some of this blame can be put on those who believe they are doing the situation good, when in reality, they are only thinking of themselves.

Britain's shrinking woodlands, often carved and sliced by man's so-called progression, are already enough of a disturbance for a population of exotic cats that now have to deal with poorly informed members of the public, hunters and in some cases trigger-happy landowners. The 'big cats' of Britain require protection, but many of the leading authorities that should be dealing with these issues have ignored the evidence for many years, so there is no point harassing such groups. The British exotic cat situation is not a modern thing, but only over the last thirty years has it really become a phenomenon, with the phases that brought us the 'beasts' of Exmoor and Bodmin, the awfully dubbed 'Fen Tiger' and other so-called 'mystery', although not at all mysterious, cats. Many sceptics still cry the question, "Where is the evidence?" despite the piles of paw-print casts, the photographs of slaughtered prey, the scratch marks twenty-feet up a tree and some reasonable photographs, and occasional snippets of camcorder footage.

Even so, when the would-be hunter obtains film of a prowling leopard or a slinking puma, what is he going to do with that footage? Well, in most cases the lure of the press and television proves too much, and whilst that researcher may have his five minutes of fame, what good is this doing to the animal and its habitat? And, if any governing body gets its paws on that film, will it be adequately monitored and produce positive action?

The elusive exotic cats that roam Britain, Australia and parts of the United States, especially the large, black cats, do not deserve to be sectioned in the folklore bracket, or classed as supernatural just because corpses are not found on a regular basis or film is not obtained. However, if the territories of these animals are methodically and exhaustively monitored then results, of a conclusive nature will be produced.

2002 has been a very productive year for KENT BIG CAT RESEARCH, having obtained conclusive video footage of various exotic species of cat, either from the public by chance or, in most cases, by monitoring areas frequently used by these beautiful animals. Of course, much of this extensive research requires patience and time and an appreciation and respect for the animals we are dealing with. Dedication and effort is obviously vital, but when reports are pieced together, encounters with these mainly nocturnal felines become quite common. The environment these animals have adapted to is ideal, encounters with the public, the cars they drive and the noises they make are common-place for these now native cats, and the age-old theories which revolve around how these cats got here are stagnant and in many cases, especially locally, can be dismissed. For now, our woodlands provide perfect shelter for felines that are often seen crossing dark lanes, rummaging through rubbish sacks in busy areas, or walking desolate railway lines. However, these cats can be monitored, filmed regularly and protected in the same way the Florida panther is, or the black leopards of Africa which are becoming less common. Unfortunately, bad press is common, despite what appears to be a keen interest from journalists etc. However, these cats are worth more than snippets, more than fleeting reports and more than those who track them.

Reports - which date back hundreds of years - are difficult to find, but they do exist. Wooded areas now would appear very small compared to the vast greenery that shrouded much of the British landscape a few centuries ago, but those that bumped into these large cats would have not reported their sightings like many members of the public do nowadays. This is not to say that many of the cats of the modern day are not the offspring of those released during the Swingin' Sixties or the mid-Seventies with the introduction of the Dangerous Wild Animals Act, but other reports and rumours of cats escaping from travelling menageries, and run-down zoos almost make great ingredients for what are becoming urban legends, in the same way as the crocodiles and alligators that appear in the New York sewers, or the out-of-place goats which turn up in New Jersey's built up areas. They are not teleported, they are not supernatural, they are not prehistoric survivors, but they have been here for many years.

So, according to so-called respected, but sceptical local zoo-keepers, "…there is no evidence to suggest that *panthera* exists in the South-East, there would be more signs." Well, let us forget the theories of government cover-up's, this is merely a situation out of hand for them, but why should those who deal with these animals be so sceptical, or, when they do admit to their existence, still believe that only one or two may be roaming the country? I feel no need to run to the sceptical folk with the film of an exotic cat. And why is that? Because at the end of the day these cats need to be left alone, and not advertised like some folklore item, or become a national celebrity in the way Bigfoot has become in parts of the U.S.A. And the petty politics already involved in the situation as people try to claw their way into the headlines is laughable, but what about the cats themselves? There will be instances when people are attacked, as well as pets. This may sound brutal but so what? These cats cannot be hunted individually, like, for instance, the case on a farm in Llangadog, in Carmarthenshire, Wales, involving a couple of large cats said to have eaten a farmers dog, causing armed police to patrol and buzz the area. The incident occurred January this year in an area known for its large cat sightings, but there will be attacks and until the authorities understand these animals, wild goose chases will be common, and these are always uncalled for, especially when a bullet could merely injure, but tragically for some humans, aggravate a large cat such as a Puma. Many who stumble through the woods looking for these animals have not a clue what to look for, or even think about how they would react if they came across a female Leopard guarding cubs, or cornered such a cat. There have already been messy situations with wild boar, another local animal still not accepted as native, despite several attacks on dogs where so-called hunters and trackers have barged into their territory.

The human race, which continues to batter the forests and fields around us, wants no contender in the countryside. The idea of large cats roaming the countryside brings back memories of the barbaric days when the majestic wolves were slaughtered, and to the modern day where toffee-nosed folk hunt foxes across the country for an ego boost. Egos may well be at work too where some exotic cat populations are concerned, as regards to cats which are illegally imported, hunted and shot for money. Kent Big Cat Research (KBCR) has already been made aware of several cases where landowners have obtained cats such as lynx, tagged them and released them in order to be hunted for huge prices.

It is not difficult to monitor local populations of large cats, despite their elusive nature many are territorial to certain areas, but yes, others are able to travel vast distances too. Males locally monitored can have a territory of up to seventy-square miles, and at times drift into other counties such as Surrey and Sussex, but there is no evidence to suggest that most of the cats in question are! The public is dealt with on a daily basis by KBCR hence the phenomenal response regarding eyewitness reports and evidence, trust and friendship is vital when dealing with witnesses or those who own land where these felines inhabit. Routes can be analysed, prey monitored and territories easily mapped out to work out just how many different cats are out there.

If cats across Britain are monitored by one particular body, and those that attempt to research them are dedicated enough, they should, on average receive around several hundred reports a month, considering KBCR is able to receive, on average, up to thirty eye-witness reports in just a few days, depending on appeals, coverage and time to survey an area where the cat in question is moving.

Most felines are moving at night, mainly due to the commotion of busy roads, which they are unable to cross during the day, but daylight reports are just as frequent as those during darkness.

If areas are monitored correctly then filming these felines and catching a glimpse of them is not as difficult as it sounds. There is no need to crash through woodland, or to set up cameras in the hope of a random visit; in fact, there is no need to invade these territories at all. Night vision equipment and other gadgets are always helpful but it is the general public that are of most use, and if they are reached enough, then it is very easy to look at a territory of a large cat, and then studying food source, boundaries these cats may face and areas they use for navigation, which will often include railway lines (constant supply of rabbits and ideal for navigation), streams and rivers (drink and navigation), golf courses, tracks, hedgerows, and mere gaps in tree-lines which these cats will use time and time again.

THE LYNX
(see Figure 1)

During August 2002, Kent Big Cat Research was able to not only see a Lynx but also film it clearly. Much footage of large cats has been taken over the years, and goodness how many people have in their possession excellent footage of animals that they do not want the media to

know about, which is good for them, and certainly keeps disturbance away from areas which these cats inhabit. Unfortunately, a lot of film from around Britain shows nothing more than dogs, and, quite bizarrely domestic cats which, in reality are nothing like their large relatives, even at a distance, but other films are often fogged by haze, darkness or ruined by shaky hand syndrome! - something FROM which the famous 1968 Patterson Bigfoot film suffers.

A contact of ours had been surveying a remote area in the county for a few weeks when he snapped a dubious picture of what appeared to be a Eurasian lynx (*lynx lynx*). The picture was taken on 6th August at 8:45 pm by the male witness who had been watching the hedgerows of a field for over an hour after recent sightings of a black leopard. Bizarrely, whilst scanning a tree line, some one hundred yards away, whilst concealing himself in undergrowth on the horizon, our contact picked out a peculiar form hiding in the shadows on this bright evening. Birds in the nearby trees had been spooked by something fifteen minutes earlier, which caused the witness to be rather curious about the area. The animal in question had begun to back away into the trees as our contact saw it, but was still evident due to the fact that a light cornfield behind it enabled our man to see its form. He took one photograph, on full optical zoom, 10X, using a Canon MV450 digital camera, before the feline form scrambled away out of sight and disappeared into dense undergrowth. Like so many other pieces of footage or still film, the cat in this photo was certainly there, but only as a blurred image, spotted on its front legs, its breast whitish, the rest of its coat mottled but its head not in view at all. Although disappointed by the picture we decided it best to monitor the area without causing a disturbance. We believed the animal would be back and was probably quite familiar with the area, as this was not a mere sighting of an animal crossing the road, but of an animal very much settled in its environment.

I and another person visited the area a few days after and sprayed the tree line with various pungent oils in the hope that the mystery cat would return. It did.

A few days after the debateable photograph, our man in the area caught the elusive and once dubious creature on film, in all its glory. On Saturday 10th August at the same, exact time, the man had taken his dog and his wife to the area, it was a sunny evening as he sat on the horizon again, his wife had taken the dog for a walk and he had settled behind the rise of a hill, looking downwards toward the tree line. After a short while, and intense scanning of the hedgerows with the naked eye and binoculars, he picked out something with his camcorder that he could not with his naked eye. Standing between two spindly trees, backing onto the same cornfield was a cat, grey in colour with tufted ears, an animal so still that it appeared ghost-like and eerie. The witness began to film and what he got was a few minutes, crystal clear, of a feline that never moved, and never looked at him either, despite the fact he was only just over one hundred yards away. He believed this was due to the fact that in the distance a cyclist flitted between another tree line and this would explain why the cat was staring in that direction - or the cat had seen his dog and remained motionless.

Our contact remained focused on the animal as it stood in the undergrowth, the top half of its body clearly visible, its face appearing almost squashed and fluffy, the animal vigilant and seemingly not at all bothered by the presence of the man. The man's dog did not sense the cat, and for the few minutes he filmed the animal, the cat never moved.

Whilst Eurasian lynx may never have fully died out and remained on this island of ours, the cat in the video appears to be Canada lynx, either deliberately released into the wilds for game or existing as an escaped feline from a private collection, or at least an offspring from cats released over a decade or so ago. The Canadian lynx (*lynx Canadensis*) can live for up to fifteen years in the wild, measure up to three-feet in length, stand up to two-and-a-half feet at the shoulder, and weigh over 10kg. Yet Canada lynx are smaller than the European lynx, which can weigh up to 20kg. Canadian lynx are long in the hind legs, have a relatively short tail, up to fifteen centimetres, and have long, black ear tufts and large feet, mainly for the snow-laden ground in their native land.

The animal in the video appears to be completely grey, is dark around the lips and shows black ear tufts. The feline stands roughly two-feet in height and probably measures around two-and-a half feet in length. After the few minutes of filming the cameraman decides to descend the hill and get a closer look, the cat appears not in the least bit phased for a while but as the witness settles the cat has turned and faces away from the man, its rear facing him, the short-bobbed tail evident, and its ears seen over its hind quarters, and then it is gone. Despite a search of the area afterwards, there was no trace of the cat, it had a whole cornfield and various thickets to vanish into.

I have viewed the film several times and am convinced the animal in the tree line is a Canadian lynx, not an overly large one, but still an exotic cat that no doubt feeds off the local birds, rabbits and rodents that live in the bushy hedges and road side dykes. Whilst the original film appears to show a lighter coloured, possibly mottled cat, which could be a Eurasian lynx, the clearer film suggests otherwise. On Sunday August 11th I would get an even better glimpse of the cat.

We set up a lookout on the Sunday from around 6:30 pm, so survey the area, to look for any tracks and pathways this sort of animal could take and use within its route. Foxes were observed and filmed as we settled. I, and the cameraman, took a position on the hillside to the right of the tree line where he had filmed the cat, and another person, armed with binoculars, stood head-on to the area where the cat had stood. We hoped that maybe it would show itself if it had intended on staying in the area, it may well have been drawn to the area originally for shelter from the sun, or food, but we believed the oils sprayed on leaves and branches would still be strong enough to attract it once more.

As the light faded and the area almost became suspended in dusk, I, and the cameraman, caught sight of the graceful feline. All three of us had been observing the tree line and the cornfield behind it, but on this occasion, at 8:50 pm, the cat had come from the extreme right, another, more sparse tree line backed by a larger cornfield joined to a small wood. Something caught our eye to the right, the large, rough, yet green field in front of us had become hazed by the dusk, and it was across this large field that the lynx had hurried across. We caught it half-way across the field, it had already come over one hundred yards without us seeing it: it was heading towards the same tree line where it had been filmed the night before. Its grey coat was not overly distinct against the field greyed by the smoky air and evening mist. It loped majestically but at great speed across the open ground, but maybe it had sensed the other watchman standing face on to its hideaway, as all of a sudden it veered to its right and headed towards an overgrown track, where it slowed but still with purpose, slinked into the undergrowth out of sight. I, and the man who had filmed it the night before, only observed the cat, and despite our signals to the other man present, he never saw a thing.

The cat was in view for what was only around five to ten seconds, yet we got a good view of its bobbed tail and again. I was one hundred percent sure it was a Canadian lynx. I believe it had been watching us for a while from the extreme right, the sparse tree line. It had obviously wanted to reach its destination, probably the same patch where it had been filmed standing, and it had a made a hasty run for it but changed its mind and direction for some reason as it neared the area. We remained in our positions for a while afterwards but to no avail and extensive searches revealed no cat, because, these are elusive animals as we know.

The lynx we saw that evening was only slightly bigger than a fox, but it moved with great speed and vanished into the air like a ghost. It enabled us, however, to monitor its territory, and since then our contact has had several glimpses and shot a few interesting pieces of film which show, in the distance, an eerie grey shape of a feline peering from the tree line from which was the area we believed the cat had run from on the 11th. A month or so after I watched the animal through binoculars for over an hour as it sat in the same thin tree line, it was on its haunches, its head low, and was eventually enveloped by the darkness of the evening. It was a privilege to witness such a solitary, shy animal, but wondered whether someone had let it go in order to hunt it or if it had been just another native cat, the product of private collections released, or escaped into the wilds many years ago.

THE BLACK LEOPARD

We were given film in 2002 of a black leopard, an exciting piece of footage shot by chance during September 2000 by a Kent man setting up CCTV camera in the bedroom of his home. The man, in his thirties, contacted me because, at the time he'd been living in a quiet, semi-rural area and had received hassle from neighbours, and so decided to set up a security camera in the hope of catching the irritants at work.

Let us go back to the night of Sunday 24th September 2000. It was a still night when Simon Whiley (name altered) had been setting up the CCTV in the bedroom. It overlooked his front garden in the seemingly quiet cul-de-sac, his bedroom provided a good view of the next few hundred yards up the road, houses to the left and right, cars parked on the flat kerbs and a nice clear night to keep an eye out. Simon had only had the camera set up for twenty minutes or so when, at 12:12 am something appeared at the top of the screen, something very weird. Around two hundred yards away to the right, near the house on the corner, a black shadow slinked eerily from the right to the left; it came from the front of the house, walked a few feet and stopped in front of a bush planted in the garden of the house in question. From where Simon was sitting, the animal he could see had stopped in front of the bush, and was doing something before it then turned around and headed back from whence it came and was gone from view. Simon couldn't believe what he'd seen and filmed, so he waited, hoping it, whatever it was, would return. For the next few hours Simon saw other animals, which helped Kent Big Cat

Research for size comparisons, but the strange creature never came back.

The film shows an animal that is only in view for a few seconds. The tape slowed shows a cat, over three feet in length, with an arched back and a very long, curved tail, powerful in the shoulder possibly and very dark in colour, emerge from the front of some houses and walk, with purpose towards a large, bush roughly three feet high and wide. The cat proceeds to linger in front of it for a few seconds, then it turns and after a few more seconds, returns from where it originally came. It is when it heads back that its shape can be observed more clearly, as there is only grey pavement and road behind it, although the cat is quite a distance from the bedroom it was being viewed from. The animal walks between the bush and the corner of the house, a gap of around four feet, and the cat fills most of this space, its hind quarters raised in comparison to its low back. It slinks away, its tail swooping down behind it, hooked to the floor.

I believe the cat had urinated on the bush to mark its territory or to attract a mate, and without a shadow of a doubt was a black leopard, a cat seen in the area many times, and one of a handful of melanistic leopards roaming the fields and woodlands. The cat is possibly a female, standing only around twenty-inches at the shoulder, but still a beautiful cat. During the Summer of 2002 we had over thirty reports in three days of two different black leopards using parallel territories, sheep kills were the mark of the cat, daylight sightings were common and this species of cat has been in this specific area for many years, long before the Swingin' Sixties when it was the 'in-thing' to parade these exotics around town, and long before the twentieth century was even a whisper, but for how long have they been native, instead of mere escapees or released pets?

The cat on the camera is a black leopard, but for anyone who would not have known this, or viewed the tape and been unconvinced, would have had their arm twisted by the appearance of a domestic feline some minutes after the main cat. The domestic cat appears only a few yards from the bedroom, leaps from a wall and on its tiny toes hurriedly scampers by. This cat is also black and appears miniature in comparison to the visitor that had emerged several minutes before and at a greater distance. A few hours after the cats had made their entrance, a fox also crosses in front of the bedroom, its relatively light coat greyed on the black and white screen, its bushy tail clear to the eye. There can be no mistaking a fox, domestic cat or any dog for a black leopard. The animal on the film has large shoulders and a very long tail, unlike any animal native to this country.

These animals have prowled into towns for years, not only to mark their ground but to raid bin bags, steal foxes and cats, and to get from A to B, and so rarely have they been seen, let alone filmed. Much of their existence can only be monitored by pure chance encounters with the public, and as these meetings with the public become more and more common, there is still no evidence to suggest that they are a threat to us. They are elusive yet curious felines, but the leopards and pumas have greater territories than the jungle cats or the caracal which can find an abundance of food in small hedgerows, but the larger cats, which can reach up to seven feet or more in length, they comb wide areas, they cross motorways at night, they prowl railway lines and back yards looking for easy meals, and despite being nocturnal and dark coloured, in the case of the melanistic Leopard, they are the most commonly sighted feline across Britain, and the population is thriving, despite these animals being rare in countries of origin.

The deep sawing cough of the black leopard has been heard across the Kent valleys for years, casts of their paw-prints have been taken, their kills examined exhaustively, yet despite being very much flesh and blood creatures, they are mystical in their appearance, I can vouch for this with the two sightings I had of one individual during Spring of 2000 (see Animals & Men 21 pp 16-21) but let us not bring such prowlers into black dog-type scenarios. These cats have merely adapted to this environment, they are seen an awful lot, but also remain unseen even more! For every five reports received there must be another five not reported, if not more. Despite the concrete intrusions upon the rural abodes, there is still much habitat for these animals to make their home, and so many people who remain blind to their existence.

A JUNGLE CAT?

It was the afternoon of July 9th 2001 when the two men set off through the Kentish country lanes. They were out to make their own historical documentary about the county, by visiting churches and scouring the deep cut lanes for signs of history. One of the men, D.B., took his car, whilst his good friend D.J. brought the camcorder. For 2 o'clock in the afternoon it was quite a breezy day, the greenery leered from the high road side like imposing walls, the roads too tight almost for one car, they hoped that on this rural ride they wouldn't meet any other vehicles: there wasn't enough room to swing a cat!

Various visits to old sites, and several strolls through the rolling fields enabled the two bumbling characters to film most of the area, but not as informatively as they would have liked. At around 2:20 pm they got back in their car and headed off through the clinging lanes, heavy woodland stretching and looming to the left, entangled bushes to their right, a tranquil, undisturbed journey through untouched Kentish countryside, only occasionally wrenched from its silence by the cackle of farmyard geese and the slow rumble of the car.

They took a claustrophobic lane into the heart of the fields, a road around twelve-feet in width, flanked by extreme greenery, the whir of the video camera punctured every now and then by D.J.'s tourist guide expressions.

"Whoa! That's done it!" D.J. exclaimed on the camcorder as another vehicle came in the other direction - the tightness of the situation and sudden confrontation causing D.J. to all of a sudden turn his camera off. A second later and the tape is running again and D.J. asks with surprise, "What's that? It's a dog…no, it's a cat."

As D.J. had got the camera going again to film the old roads of Kent, something - something unusual - moved across the road fifty yards ahead and in a couple of seconds had disappeared into the woods. For some reason, the two men in the car did not bother to search the area where the creature had vanished into; instead, they carried on filming their historical feature until a few weeks afterwards, when they heard about Kent Big Cat Research in a newspaper.

What the men filmed we have analysed thoroughly and it shows, albeit it for a very brief moment, a cat, midway across the lane, heading from right to left slightly in the direction of the car. Although the vehicle approaches at reasonable speed, the car is in no hurry at all to reach the grass verge and as the cameraman finally zooms, it strolls into the woods out of sight. The animal is lean, long-legged and sandy-coloured, its ears are long, but not lynx-like or marked, the body is not marked, but pale all over and a paler underside to the body. The cat stands around fifteen inches at the shoulder and measures over two-feet in length and has a tail, not overly long like that of a puma, and it is tipped black. As the cat walks into the undergrowth its ears are alertly pointed outwards to its flanks.

The cat on the film is a jungle cat, also known as a reed cat (*Felis chaus*). Its distribution extends from Egypt, through Jordan, the Middle East, and further east into India, Sri-Lanka and south-western China. The coloration of the jungle cat, like so many other cats, varies greatly, in this instance from sandy-yellow to greyish to reddish, having brown stripes on the legs and similar rings on the tail and a black tip to the tail, and the ears are also tipped with black, tall and rounded, and they rely very much on their hearing to locate small prey such as mice and insects, and these cats are often on the move during daylight. In his book '*Wild Cats Of The World*', author C.A.W. Guggisberg observes that, "...Jungle Cats usually move at a slow, careful and noiseless trot..... it does not shun the vicinity of man."

The measurements and descriptions he gives for the cat match those in the video, which merely proves what a variety of feline species roam Britain, and cats such as jungle cats, ocelots and the caracal can remain undetected for years, and even those that are sighted, will not be reported because witnesses do not know what they are seeing, especially if encounters are as brief as the one caught on this video.

A majority of the cats that roam Britain, if they avoid natural diseases or viruses such as cat flu, will live out their natural lives. In their countries of origin human persecution is obliterating some species of cats, and that is likely here if they are not accepted as native animals and are forever considered as monsters. At the moment large cats such as the black leopard remain seated on the folklore fence, and maybe it is best if they stay that way considering authorities that have the power to help them, but obviously do not realise it, are ignoring their existence. Evidence has mounted for many decades to their existence and the pieces of film spoken of here are nothing more than extra ingredients to a situation that could, if ignored for too long, take on a snowball effect. As the situation stands, a few of these cats could be killed, whether on roads or by bullet, and all the while they are making the wrong headlines, mainly due to the press who can turn an interesting story of an escaped ocelot into a, "…beast on the loose" yarn, there will always be some maniac prowling the fields with their gun, or more and more occasions where armed police are sent on a mission to destroy the 'beast of the moors', and not once are these cats considered or understood.

Farmers will lose livestock and in some parts of the world they are compensated for their losses, but over here they know these cats exist but they are getting no answers, or understanding of these animals, and it is this ignorance which needs to be eradicated so that these felines, that have been here for hundreds of years, may elude eradication themselves. There may be many members of the public who are opposed to the situation where exotic cats are accepted as British wildlife, and there are still those who do not believe in their existence anyway, but for how much longer can the scare stories continue to paint deceptive pictures of animals that are only killing to survive, just like they have done for many years in Canada, Africa, Asia etc, etc? If the cats inhabiting Britain are shot and killed, it becomes a dramatic news story, when the reality is, it should be regarded as a tragedy.

The case of the creeping fox terrier clone
or
how a sea serpent sighting crept back in time.

by Chris. M. Moiser

Back in the late 1980s I used to subscribe to the American magazine *Natural History*. This was (and still is) an impressive magazine published by the American Museum of Natural History. One of its selling points, as far as I was concerned, was, every month, an essay by Stephen Jay Gould, the evolutionary biologist. These essays had titles such as "Bully for Brontosaurus", "Kropotkin was no crackpot", and "Male nipples and clitoral ripples". Each essay made a point about some aspect of evolutionary biology and combined it with an interesting glimpse of a facet of the history of science. It was learning at its best, a series of interesting, but apparently unconnected, strange facts, generally supporting mainstream Darwinian theory.

One particularly impressive essay was "The Case of the Creeping Fox Terrier Clone" which described the tendency for published mistakes to continue to appear in textbooks after they have been recognised or after they should have been recognised. He described an original paper by Diane Paul in *The Sciences* in May 1987. Paul had looked at twenty-eight textbooks on introductory genetics published between 1978 and 1984. She particularly examined the way in which the (totally discredited) work of Sir Cyril Burt had been incorporated. Burt had invented his data, his co-workers and his results; it is difficult to imagine a way of bringing greater discredit on one's work other than to do that! Despite the fact that several of the books carried, as warning, details of the Burt scandal, nearly half of them included his data. Clearly this had been done unconsciously. Ten of the twenty-eight textbooks even included a figure from a review article in Science in 1963, which featured Burt's results. (Burt's results were not, of course, suspect in 1963). Paul had suggested that the authors of these textbooks had not read the original texts and had copied from other textbooks in writing their own. There is no other easy explanation for a situation where the author gives a warning about the Burt scandal and then goes on and uses the discredited data in a figure.

Paul had suggested that the increased commercialisation of textbooks led to this virtual cloning of contents with the actual text of the book being secondary to the presentation, and subsidiary materials in the form of teachers guides, slide sets, etc. Gould in agreeing with this suggested that market forces in publishing have encouraged this tendency to "clone" data from other texts, particularly in areas where the author has no direct personal knowledge certain areas of the subject matter.

Gould then continues to look at one of the standard textbook examples of a fossil record of evolution, that of the horse. In particular he considers *Eohippus* (dawn horse), or as it is more properly known, *Hyracotherium*. Most of the modern books describe this animal as being the size of a "fox-terrier". In fact Owen (yes, Richard Owen, who invented the word "dinosaur" and was possibly Victorian England's most prestigious palaeontologist) first described the animal in 1841 as being either hare-like, or being something between a hyrax and a hog. Most subsequent writers in the nineteenth century described it as being like a fox, and it was only during the twentieth century that the simile of a fox terrier was given. Gould had, when watching the "Thin Man" films seen *Asta*, a fox terrier, but not really registering its size at the time. (Had Gould read the Dashiell Hammett book "The Thin Man" rather than having seen the film, he would have known that the original, *Asta* was in fact a Schnauzer, and not a fox terrier). In fact most modern estimates of the size of *Hyracotherium* put its weight at something like 25 kilograms or 55 pounds. This suggests that *Hyracotherium* was considerably larger than the fox terrier. A commonly copied comparison is therefore an incorrect one too.

The point of Gould's essay is one of identifying the evils of what he calls "textbook cloning", he suggests that it is almost a substitute for thinking and striving to improve. He believed that a "carelessly cloned work" would not excite students, however "state-of-the-art" the production of the book is. In the United Kingdom our "National Curriculum" and the reduction in exam boards must provide an even greater incentive to "clone text", if only to permit greater time to be spent on developing the associated CD Rom, with "end of chapter tests" and an interactive web site.

My own example is rather more "crypto", but it also makes a second point, which my critics will probably label as self-aggrandisement if there is a suggestion that I am attempting to improve on Gould, which I am not. The point is very simply that when you find an original reference, particularly a historic one, there is often other contemporary evidence to be discovered, whether it be circumstantial or direct evidence. Sadly such rewards are rarely exploited to their full advantage.

I am, at present, in my spare time, attempting to write a small book on the sea serpents of the South West. Examples of local sightings are relatively easy to find for the twentieth and twenty-first century, but those for the nineteenth century are very scarce. Do not misunderstand me, there are many, many nineteenth century sightings of such animals, but the majorities seem to be off the American East coast, and in other exotic locations. The South coast of England was either very short of them or they just did not get reported at the time.

Most of the classic texts such as Heuvelmans [1968] and Bright [1989] give two examples from the South West, the 1876 report from the *West Briton* and the 1882 report of the sighting at Bude by the Reverend E. Highton. The reference that is given for the latter sighting is *The Times* of the 12th of October 1882. When I saw this reference I decided to have a look at it. The local library has a full set of microfilm archives of *The Times*. It is possible to photocopy the articles that you want, and providing that the machines are not under pressure, to spend time perusing the papers. If you are writing articles to a close editorial brief this can give you a few words as extra filling to reach the required word count. I remember once "filling out" an article on travelling menageries by pointing out that whilst the said menagerie was in town there was much correspondence nationally over whether Mr. Stanley would find Dr. Livingstone in Africa. Possibly not too relevant to the article, but it does put it in historical perspective. The other advantage is that if one finds the report it is then possible to scan the newspapers for the following week to see if the report has elicited further correspondence or reports. In one case I found further eyewitness accounts of the original sighting from a totally different position. This was good confirmation of the accuracy of the report, and made it even more believable because the two parties apparently did not know each other.

Where the reference is in a national publication and can be found and confirmed it is always then a good idea to view any local papers of the same date. Even if they do not carry the sighting they may carry information, which contributes to the overall view. In the case of sea serpent reports the local weather may be very relevant, as may the state of the local fishing industry.

I have conducted this type of "contemporary scanning" for years, often with interesting and sometimes unexpected results. The problem was that with the Reverend Highton's sighting, which was supposed to have been reported in the Times of 12th October 1882 I just couldn't find it. I read the newspapers for 6 days either side of the report, and still couldn't find it. At this stage it was becoming a matter of honour, so I telephoned *The Times* and spoke to the records section. A lovely, lovely lady, who I think was called Tamsin, spoke to me, seemed interested in the problem, and said that she would get back to me. She kept her word, and within 48 hours there was a message on my answerphone. Sadly further attempts to contact Tamsin to thank her failed. The actual date of the reference was 17th October 1883, over a year after the (often) quoted one. A quick trip back to the local library confirmed the date, and 30p acquired me a copy of the reverend Highton's letter. There was more information in his report than the classic sea serpent books carried books carried, but not a lot more.

The normal scan of the next few days' papers also came up with a find. On the 20th of October 1883 Vice Admiral W. Gore Jones recorded his experience of a sea serpent. In 1848 he was on HMS St. Vincent which was lying off Spithead (Portsmouth), it was a summer evening and dinner was about to be served in the officer's mess when the midshipman off the watch came in to announce that a sea serpent had been sighted. Dinner was abandoned and glasses fetched, a shaggy maned sea serpent was seen about a mile off. Ship's boats were launched with some of the officers having fetched their guns, intent on shooting the mystery animal. Gore Jones stayed on H.M.S. St. Vincent but reports that as the ships boats got close to the "serpent" all attempts to shoot it stopped and there was much laughter. The "serpent" was in fact a long line of soot. This was presumed to have come from a steamer in Southampton water having cleaned its chimneystack or other "tubes". The soot had stuck together and was carried out on the tide as a long slick. This explains his sighting of a sea serpent and he suggests that this might be what the Reverend Highton had seen a few days earlier. The sea off Bude is of course different to that in Southampton Water and large quantities of soot masquerading as sea serpents off North Cornwall seem unlikely because of the less sheltered position.

During my initial search for the original Bude report I did in fact consult a number of textbooks that are accepted authorities on sea-serpents, partly checking for other references, partly checking to see whether there were any other clues as to where I might find more details of the Reverend Highton's sighting. All gave the same reference. It seems that the original reference was that early classic *"The Great Sea-Serpent"* by A.C. Oudemans, first published in the UK in 1892. It then appears in Heuvelmans, *"In the Wake of the Sea-serpents"* in 1968, and is subsequently quoted by Michael Bright in *"There are Giants in the Sea"* in 1989. Several lesser known texts also carried the (erroneous) 1882 reference. The most recent major sea serpent text, Paul Harrison's, *"Sea Serpents and Lake Monsters of the British Isles,* published in 2001, has actually hybridised the dates. He gives the date in *The Times* as 17th October 1882, but then quotes the letter in full, giving the dates of 11th and 12th October 1883 in the letter!

It would appear that all of the incorrect references can be traced back to Oudemans. On closer examination Oudemans actually credits RPG with the reference. RPG was a gentleman who had sent Oudemans a considerable number of reports, which Oudemans gratefully acknowledged. We will never know whether Oudemans had cause to believe that one or two of the references would be inaccurate and was distancing himself from them or whether, as seems more likely, he was being the proper scientist and acknowledging all his sources correctly.

There are no shortcuts to decent research, even in this age of the broadband internet and high-speed connection, references still need to be checked. Checking old references does, in the main need library time, but when going through those archives one, of course, often explores areas that may not have received academic scrutiny for a long time, and possibly never by a researcher with modern insight into the subject area. The message is that if you do make a discovery just be careful to record the date correctly, and if you do debunk a major myth do proof read the rest of your text to be certain that you are not still confirming it elsewhere. This is serious science and cryptozoology should now be treating itself as a serious science and setting itself the highest standards.

References

Bright, M. (1989) *There are Giants in the Sea* Robson Books, London.
Gould, S.J. (1991) *Bully for Brontasaurus-Reflections in Natural History,* Hutchinson Radius, London.
Harrison, P. (2001) *Sea Serpents and Lake Monsters of the British Isles,* Robert Hale Ltd., London.
Heuvelmans, B. (1968) *In the Wake of the Sea-serpents,* Rupert Hart-Davis, London.
Oudemans, A.C. (1892) *The Great Sea-serpent. An Historical and Critical Treatise.* Luzac and Co. (London)

Postscript

A further example of the multiple lives of quoted misinformation was drawn to my attention by the learned editor of this yearbook as I was finishing the article. In fact it isn't so much quoted misinformation as a selective quoting, out of context, of a report of an attack on two boys by a large bird between Mount Hawke and Porthowan in Cornwall. The original report was in the *Cornish Echo* of June 4th, 1926. It was possibly first quoted out of context by A. Mawnan-Peller (nom-de-plume, if ever there was one), in his book *Morgawr the Monster of Falmouth Bay.*

Here the quotation suggests that the strange bird that attacked the boys might be an ancestor of "The Owlman of Mawnan" - a mystical half man half owl first seen in the Mawnan area in the 1970s. By the *ejusden generis* rule this suggests some large bird of prey. Janet and Cohn Bord, in their book *Alien Animals* (1980) place the report in the middle of a chapter on "Giant birds and birdmen" again implying a large carnivorous bird. In *"The Owlman and others"* Jonathan Downes actually gives the whole report, and prevents the cryptozoological Chinese whispers game from creating another totally unidentifiable monster. Although not able to identify the bird, the description gives it a wingspan of 6 foot 3 inches and a body length of three feet. Additionally it has webbed feet and a duck-shaped body. This immediately contradicts any suggestion of it being a bird of prey and is strongly suggestive of it being some type of waterfowl about the size of the Canada goose.

DARWINISM:
A Crumbling Theory

An overlooked explanation for why the fossil record shows primitive and complex life appearing suddenly on Earth, with no predecessors, is extraterrestrial intervention.

by Lloyd Pye

PART ONE

Since writing my first essay for NEXUS in mid-2002 [see 9/04], I've been bombarded by emails (nearing 200) from around the world, many offering congratulations (always appreciated, of course) and many others requesting more instruction or deeper insight into areas discussed and/or not discussed.

Let's face it: nearly everyone is interested in Darwinism, Creationism, Intelligent Design, and the new kid in town, Interventionism. Because of length constraints, this essay must be in two parts. Here, in Part One, I'll go over the basics currently known about the origin of life on Earth. Later, in Part Two, I'll discuss what is known and what can be safely surmised about the origin of humanity.

We begin by understanding that Charles Darwin stood on a very slippery slope when trying to explain how something as biologically and biochemically complex as even the simplest form of life could have spontaneously generated itself from organic molecules and compounds loose in the early Earth's environment. Because that part of Darwin's theory has always been glaringly specious, modern Darwinists get hammered about it from all sides, including from the likes of me, with a net result that the edifice of "authority" they've hidden behind for 140 years is crumbling under the assault.

Imagine a mediaeval castle being pounded by huge stones flung by primitive, but cumulatively effective, catapults. Darwinism (and all that term has come to represent: natural selection, evolution, survival of the fittest, punctuated equilibrium, etc.) is the castle; Darwinists man the battlements as the lobbed stones do their work; Intelligent Designers hurl the boulders doing the most damage; Creationists, by comparison, use slings; and the relatively few (thus far) people like me, Interventionists, shoot a well-aimed arrow now and then, though nobody pays much attention to us yet.

Remember, a well-aimed (or lucky - in either case, the example is instructive) arrow took down mighty Achilles. Darwinists have heels, too.

LIFE, OR SOMETHING LIKE IT

In Charles Darwin's time, nothing was known about life at the cellular level. Protoplasm was the smallest unit they understood. Yet Darwin's theory of natural selection stated that *all* of life - every living entity known then or to be discovered in the future - simply *had* to function from birth to death by "natural laws" that could be defined and analysed. This would of course include the origin of life. Darwin suggested life might have gradually assembled itself from stray parts lying about in some "warm pond" when the planet had cooled enough to make such an assemblage possible. Later it was realised that nothing would likely have taken shape (gradually or otherwise) in a static environment, so a catalytic element was added: lightning.

Throughout history up to the present moment, scientists have been forced to spend their working lives with the "God" of the Creationists hovering over every move they make, every mistake, every error in judgment, every personal peccadillo. So when faced with something they can't explain in rational terms, the only alternative option is "God did it", which for them is unacceptable. So they're forced by relentless Creationist pressure to come up with answers for absolutely everything that, no matter how absurd, are "natural". That was their motivation for the theory that a lightning bolt could strike countless random molecules in a warm pond and somehow transform them into the first living creature. The "natural" forces of biology, chemistry and electromagnetism could magically be swirled together - and *voilà*! - an event suspiciously close to a miracle.

Needless to say, no Darwinist would accept terms like "magic" or "miracle", which would be tantamount to agreeing with the Creationist argument that "God did it all". But in their heart-of-hearts, even the most fanatical Darwinists had to suspect the "warm pond" theory was absurd.

And as more and more was learned about the mind-boggling complexity of cellular structure and chemistry, there could be no doubt. The trenchant Fred Hoyle analogy still stands: it was as likely to be true as that a tornado could sweep through a junkyard and correctly assemble a jetliner.

Unfortunately, the "warm pond" had become a counterbalance to "God did it", so even when Darwinists knew past doubt that it was wrong, they clung to it, outwardly proclaimed it and taught it. In many places in the world, including the USA, it's still taught.

TOO HOT TO HANDLE

The next jarring bump on the Darwinist road to embattlement came when they learned that in certain places around the globe there existed remnants of what had to be the very first pieces of the Earth's crust. Those most ancient slabs of rock are called cratons, and the story of their survival for 4.0 billion [4,000,000,000] years is a miracle in itself. But what is most miraculous about them is that they contain fossils of "primitive" bacteria! Yes, bacteria, preserved in 4.0-billion-year-old cratonal rock. If that's not primitive, what is? However, it presented Darwinists with an embarrassing conundrum.

If Earth began to coalesce out of the solar system's primordial cloud of dust and gas around 4.5 billion years ago (which by then was a well-supported certainty), then at 4.0 billion years ago the proto-planet was still a seething ball of cooling magma. No warm ponds would appear on Earth for at least a billion years or more. So how to reconcile reality with the warm-pond fantasy?

There was *no* way to reconcile it, so it was ignored by all but the specialists who had to work with it on a daily basis. Every other Darwinist assumed a position as one of the "see no evil, speak no evil, hear no evil" monkeys. To say they "withheld" the new, damaging information is not true; to say it was never emphasised in the popular media for public consumption is true.

That has become the way Darwinists handle any and all challenges to their pet theories: if they can no longer defend one, they don't talk about it, or they talk about it as little as possible. If forced to talk about it, they invariably try to "kill the messenger" by challenging any critic's "credentials". If the critic lacks academic credentials equal to their own, he or she is dismissed as little more than a crackpot. If the critic has equal credentials, he or she is labelled as a "closet Creationist" and dismissed. No career scientist can speak openly and vociferously against Darwinist dogma without paying a heavy price. That is why and how people of normally good conscience can be and have been "kept in line" and kept silent in the face of egregious distortions of truth.

If that system of merciless censure weren't so solidly in place, then surely the next Darwinist stumble would have made headlines around the world as the final and absolute end to the ridiculous notion that life could possibly have assembled itself "naturally". They couldn't even be sure it happened on Earth.

TWO FOR THE PRICE OF ONE

The imposing edifice of Darwinian "origin of life" dogma rested on a piece of incontrovertible bedrock: there could be only one progenitor for *all* of life. When the fortuitous lightning bolt struck the ideally concocted warm pond, it created only *one* entity. However, it was no ordinary entity. With it came the multiple ability to take nourishment from its environment, create energy from that nourishment, expel waste created by the use of that energy and (almost as an afterthought) reproduce itself ad infinitum until one of its millions of subsequent generations sits here at this moment reading these words. Nothing miraculous about that; simply incalculable good fortune.

This was Darwinist gospel - preached and believed - until the bacteria fossils were found in the cratons. Their discovery was upsetting, but not a deathblow to the Darwinist theory. They had to concede (among themselves, of course) that the first life-form didn't assemble itself in a warm pond, but it came together somehow because every ancient fossil it spawned was a single-celled bacteria lacking a cell nucleus (*prokaryotes*). Prokaryotes preceded the much later single-celled bacteria *with* a nucleus (*eukaryotes*), so the post-craton situation stayed well within the Darwinian framework. No matter how the first life-form came into existence, it was a single unit lacking a cell nucleus, which was mandatory because even the simplest nucleus would be much too "irreducibly complex" (a favourite Intelligent Design phrase) to be created by a lightning bolt tearing through a warm pond's molecular junkyard. So the Darwinists still held half a loaf.

In the mid-1980s, however, biologist Carl Woese stunned his colleagues with a shattering discovery. There wasn't just the predicted (and essential) single source for all forms of life; there were *two*: two types of prokaryotic bacteria as distinct as apples and oranges, dogs and cats, horses and cows - two distinct forms of life, alive and well on the planet at 4.0 billion years ago. Unmistakable. Irrefutable. Get over it. Deal with it.

But how? How to explain separate forms of life springing into existence in an environment that would make hell seem like a summer resort? With nothing but cooling lava as far as an incipient eye might have seen, how could it be explained in "natural" terms? Indeed, how could it be explained in any terms other than the totally unacceptable? Life, with all its deepening mystery, had to have been *seeded* onto Earth.

PANSPERMIA RAISES ITS UGLY HEAD

Panspermia is the idea that life came to be on Earth from somewhere beyond the planet and possibly beyond the solar system. Its means of delivery is separated into two possible avenues: directed and undirected.

Undirected panspermia means that life came here entirely by accident and was delivered by a comet or meteor. Some scientists favour comets as the prime vector because they contain ice mixed with dust (comets are often referred to as "dirty snowballs"), and life is more likely to have originated in water and is more likely to survive an interstellar journey frozen. Other scientists favour asteroids as the delivery mechanism because they are more likely to have come from the body of a planet that would have contained life. A comet, they argue, is unlikely ever to have been part of a planet, and life could not possibly have generated itself in or on a frozen comet.

Directed panspermia means life was delivered to Earth by intelligent means of one kind or another. In one scenario, a capsule could have been sent here the same way we sent *Voyager* on an interstellar mission. However, if it was sent from outside the solar system, we have to wonder how the senders might have known Earth was here, or how Earth managed to get in the way of something sent randomly (*à la Voyager*).

In another scenario, interstellar craft manned by extraterrestrial beings could have arrived and delivered the two prokaryote types. This requires a level of open-mindedness that most scientists resolutely lack, so they won't accept either version of directed panspermia as even remotely possible. Instead, they cling to their "better" explanation of undirected panspermia because it allows them to continue playing the "origin" game within the first boundaries set out by Charles Darwin: undirected is "natural"; directed is "less natural".

Notice it can't be said that directed panspermia is "*un*natural". According to Darwinists, no matter where life originated, the process was *natural* from start to finish. All they have to concede is that it didn't take place on Earth. However, acknowledging that forces them to skirt dangerously close to admitting the reality of extraterrestrial life, and their ongoing "search" for such life generates millions in research funding each year. This leaves them in no hurry to make clear to the general public that, yes, beyond Earth there is at the very least the same primitive bacterial life we have here. There's no doubt about it. But, as usual, they keep the lid on this reality, not exactly hiding it but making no effort to educate the public to the notion that we are not, and never have been, alone. The warm pond still holds water, so why muddy it with facts?

A PATTERN EMERGES

In my book, *Everything You Know Is Wrong*, I discuss all points mentioned up to now, which very few people outside academic circles are aware of. Within those circles, a hard core of "true believers" still seizes on every new discovery of a chemical or organic compound found in space to try to move the argument back to Darwin's original starting point that *somehow* life assembled itself on Earth "naturally".

However, most objective scholars now accept that the first forms of life had to have been *delivered* because: (1) they appear as two groups of multiple prokaryotes (archaea and true bacteria); (2) they appear whole and complete; (3) the hellish primordial Earth is unimaginable as an incubator for burgeoning life; and (4) a half-billion years seems far too brief a time-span to permit a gradual, step-by-step assembly of the incredible complexity of prokaryotic biology and biochemistry.

Even more damaging to the hard-core Darwinist position is that the prokaryotes were - quite propitiously - as durable as life gets. They were virtually indestructible, able to live in absolutely any environment - and they've proved it by being here today, looking and behaving the same as when their ancestors were fossilised 4.0 billion years ago. Scalding heat? *We love it!* Choked by saline? *Let us at it!* Frozen solid? *We're there!* Crushing pressure? *Perfect for us!* Corrosively acidic? *Couldn't be better!*

Today they are known as *extremophiles*, and they exist alongside many other prokaryotic bacteria that thrive in milder conditions. It would appear that those milder-living prokaryotes could not have survived on primordial Earth, so how did they come to be? According to Darwinists, they "evolved" from extremophiles in the same way humans supposedly evolved on a parallel track with apes - from a "common ancestor".

Darwinists contend such parallel tracks don't need to be traceable. All that's required is a creature looking reasonably like another to establish what they consider a legitimate claim of evolutionary connection. Extremophiles clearly existed: we have their 4.0-billion-year-old fossils. Their descendants clearly exist today, along with mild-environment prokaryotes that must have descended from them. However, transitional forms between them cannot be found, even though such forms are required by the tenets of Darwinism. Faced with that embarrassing problem, Darwinists simply insist that the missing transitional species *do* exist, still hidden somewhere in the fossil record, just as the "missing link" between apes and humans is out there somewhere and will indeed be discovered someday. It's simply a matter of being in the right place at the right time.

For as expedient as the "missing link" has been, it's useless to explain the next phase of life on Earth, when prokaryotes began sharing the stage with the much larger and much more complex (but still single-celled) eukaryotes, which appear around 2.0 billion years ago. The leap from prokaryote to eukaryote is too vast even to pretend a missing evolutionary link could account for it. A dozen would be needed just to cover going from no nucleus to one that functions fully. (This, by the way, is also true of the leap between so-called pre-humans and humans, which will be discussed in Part Two).

How to explain it? Certainly not plausibly. Fortunately, Darwinists have never lacked the creativity to invent "warm-pond" scenarios to plug holes in their dogma.

DOING THE DOGMA SHUFFLE

Since it's clear that a "missing link" won't fly over the prokaryote-eukaryote chasm, why not assume some of the smaller prokaryotes were eaten by some of the larger ones? *Yeah, that might work!* But instead of turning into food, energy and waste, the small ones somehow turn themselves - or get turned into - cell nuclei for larger ones. *Sure, that's a keeper!* Since no one can yet prove it didn't happen (*Thank God!*), Darwinists are able to proclaim it did. (Keep in mind, when any critic of Darwinist dogma makes a suggestion that similarly can't be proved, it's automatically dismissed, because "lack of provability" is a death sentence outside their fraternity. Inside their fraternity, consensus is adequate because the collective agreement of so many "experts" should be accepted as gospel.)

To Interventionists like me, the notion of prokaryotes consuming each other to create eukaryotes is every bit as improbable as the divine fiat of Creationists. But even if it were a biological possibility (which most evidence weighs against), it would still seem fair to expect "transition" models somewhere along the line. Darwinists say "no" because this process could have an "overnight" aspect to it. One minute there's a large prokaryote alongside a small one, the next minute there's a small eukaryote with what appears to be a nucleus inside it. Not magic, not a miracle, just a biological process unknown today but which could have been possible 2.0 billion years ago. Who's to say, except an "expert"? In any case, large and small prokaryotes lived side by side for 2.0 billion years (long enough, one would think, to learn to do so in harmony), then suddenly a variety of eukaryotes appeared alongside them, whole and complete, ready to join them as the only game in town for another 1.4 billion years (with no apparent changes in the eukaryotes, either).

At around 600 million years ago, the first multicellular life- forms (the Ediacaran Fauna) appear - as suddenly and inexplicably as the prokaryotes and eukaryotes. To this day, the Ediacaran Fauna are not well understood, beyond the fact they were something like jellyfish or seaweeds in a wide range of sizes and shapes. (It remains unclear whether they were plants or animals, or a bizarre combination of both.) They lived alongside the prokaryotes and eukaryotes for about 50 million years, to about 550 million years ago, give or take a few million, when the so-called "Cambrian Explosion" occurred.

It's rightly called an "explosion", because within a period of only 5 to 10 million years - a mere eye-blink relative to the 3.5 billion years of life preceding it - the Earth's oceans filled with a dazzling array of seawater plants and all 26 of the animal phyla (body types) catalogued today, with no new phyla added since. No species from the Cambrian era looks like anything currently alive - except trilobites, which seem to have spawned at least horseshoe crabs. However, despite their "alien" appearance, they all arrived fully assembled - males and females, predators and prey, large and small, ready to go. As in each case before, no predecessors can be found.

THE PACE HEATS UP

Volumes have been written about the Cambrian Explosion and the menagerie of weird plants and animals resulting from it. The Earth was simply *inundated* with them, as if they'd rained down from the sky. Darwinists concede it is the greatest difficulty - among many - they confront when trying to sell the evolutionary concept of *gradualism*. There is simply no way to reconcile the breathtaking suddenness, the astounding variety and the overwhelming incongruity of the Cambrian Explosion. It is a testament to the old adage that "one ugly fact can ruin the most beautiful theory". But it's far from the only one.

All of complex life as we understand it begins with the Cambrian Explosion, in roughly the last 550 million years. During that time, the Earth has endured five major and several minor catastrophic extinction events. Now, one can quibble with how an event catastrophic enough to cause widespread extinctions could be called "minor", but when compared to the major ones the distinction is apt. The five major extinction events eliminated 50% to 90% of all species of plants and animals alive when the event occurred.

We all know about the last of those, the Cretaceous event of 65 million years ago that took out the dinosaurs and much of what else was alive at the time. But what few of us understand is the distinctive pattern to how life exists *between* extinction events and *after* extinction events. This difference in the pattern of life creates serious doubts about "gradualism" as a possible explanatory mechanism for how species proliferate.

Between extinction events, when environments are stable, life doesn't seem to change at all.

The operative term is stasis. Everything stays pretty much the same. But after extinction events, the opposite occurs: everything changes profoundly. New life-forms appear all over the place, filling every available niche in the new environments created by the after-effects of the catastrophe. Whatever that is, it's not gradualism.

In 1972, (the late) Stephen J. Gould of Harvard and Niles Eldredge of the American Museum of Natural History went ahead and bit the bullet by announcing that fact to the world. Gradual evolution simply was not borne out by the fossil record, and that fact had to be dealt with. Darwin's view of change had to be modified. It wasn't a gradual, haphazard process dictated by random, favourable mutations in genes. It was something else.

That "something else" they called punctuated equilibrium. The key to it was their open admission of the great secret that life-forms only changed in spurts after extinction events, and therefore had nothing to do with natural selection or survival of the fittest or any of the old Darwinist homilies that everyone had been brainwashed to believe. It was the first great challenge to Darwinian orthodoxy, and it was met with furious opposition. The old guard tagged it "punk eek" and called it "evolution by jerks".

TRUTH AND CONSEQUENCES

What Gould and Eldredge were admitting was the great truth that evolution by natural selection is not apparent in either the fossil record or in the life we see around us. The old guard insisted that the fossil record simply *had* to be wrong: it wasn't giving a complete picture because large tracts of it *were* missing. That was true, but much larger tracts were available, and those tracts showed the overwhelming stasis of life-forms in every era, followed by rapid filling of environmental niches after each extinction event. So while parts of the record were indeed missing, what was available was unmistakable.

Arguments raged back and forth. Explanations were created to try to counter every aspect of the punk-eek position. None was ever particularly convincing, but they began to build up. Remember, scientists have the great advantage of being considered by one and all as "experts", even when they haven't the slightest idea of what they're talking about. That allows them to throw shot after shot against the wall until something sticks, or until the target of their wrath is covered in so much "mud" that it can't be seen any more. Such was the fate of the punk-eekers. By the early 1990s, they'd been marginalised.

One can hardly blame the old-guard Darwinists for those attacks. If granted any credence, the sudden radiations of myriad new species into empty environmental niches could have gutted many of the most fundamental tenets of gradual, "natural" evolution. That idea simply could *not* become established as a fact. Why? Because the warm pond was drained dry, biochemistry was rendering the "small-eaten-by-large prokaryotes turned into eukaryotes" story absurd, and the Cambrian Explosion was flatly inexplicable. If "sudden radiation" were heaped onto all of that, the entire theory of evolution could flounder, and where would that leave Darwinists? Facing righteous Creationists shouting, "See! God *did* do it after all!" Whatever else the Darwinists did, they couldn't allow that to happen.

Speaking as an Interventionist, I don't blame them. To me, God stands on equal footing with the lightning bolt. I see a better, far more rational answer to the mysteries of how life came to be on planet Earth: it was put here by intelligent beings, and it has been continuously monitored by those same beings. Whether it's been developed for a purpose or toward a goal of some kind seems beyond knowing at present, but it can be established with facts and with data that intervention by outside intelligence presents the most logical and most believable answer to the question of how life came to be here, as well as of how and why it has developed in so many unusual ways in the past 550 million years.

So now we come to the crux.

COSMIC ARKS

Darwinists go through life waving their PhD credentials like teacher's pets with a hall pass, because it allows them to shout down and ridicule off the public stage anyone who chooses to avoid the years of brainwashing they had to endure to obtain those passes. However, their credentials give them "influence" and "credibility" with the mainstream media, who don't have the time, the ability or the resources to make certain that everything every Darwinist says is true. They must trust *all* scientists not to have political or moral agendas, and not to distort the truth to suit those agendas. So, over time, the media have become lapdogs to the teacher's pets, recording and reporting whatever they're told to report, while dismissing out of hand whatever they're told to dismiss out of hand.

Despite Darwinists' rants that those who challenge them do so out of blithering idiocy, that is not always the case. For that matter, their opponents are not all Creationists, or even Intelligent Designers, whom Darwinists labour feverishly to paint into the "goofy" corner where Creationists rightly reside. So Interventionists like me have few outlets for our ideas, and virtually none in the mainstream media. Nevertheless, we feel our view of the origin of life makes the best sense, given the facts as they are now known, and the most basic aspect of our view starts with what I once called "cosmic dump trucks". However, that term has been justly criticised as facetious, so now I call them "cosmic arks".

Imagine this scenario: a fleet of intergalactic "terraformers" (another term I favour) cruises the universe. Their job is to locate forming solar systems and seed everything in them with an array of basic, durable life-forms capable of living in *any* environment, no matter how scabrous. Then the terraformers return on a regular basis, doing whatever is needed to maximise the capacity for life within the developing solar system. Each system is unique, calling for specialised forms of life at different times during its development, which the terraformers provide from a wide array of cosmic arks at their disposal.

With that as a given, let's consider what's happened on Earth. Soon after it began to coalesce out of dust and gas, two forms of virtually indestructible bacteria appeared on it, as if someone knew precisely what to deliver and when.

Also, it would make sense that every other proto-planet in the solar system would be seeded at the same time. How could even terraformers know which forming planets would, after billions of years, become habitable for complex life? And guess what? A meteorite from Mars seems to contain fossilised evidence of the same kinds of *nano-* (extremely small) bacteria found on

Earth today. All other planets, if they're ever examined, will probably reveal similar evidence of a primordial seeding. It would make no sense for terraformers to do otherwise.

THE RUST ALSO RISES

So, okay, our solar system is noticed by intergalactic terraformers as the new sun ignites and planets start forming around it. On each of the planets they sprinkle a variety of two separate forms of single-celled bacteria they know will thrive in *any* environment (the extremophiles). But the bacteria have a purpose: to produce oxygen as a component of their metabolism. Why? Because life almost certainly has the same basic components and functions everywhere in the universe. DNA will be its basis, and "higher" organisms will require oxygen to fuel their metabolism. Therefore, complex life can't be "inserted" anywhere until a certain level of oxygen exists in a planet's atmosphere.

Wherever this process is undertaken, the terraformers have a major problem to deal with: *iron*. Iron is an abundant element in the universe. It is certainly abundant in planets (meteorites are often loaded with it). Iron is very reactive with oxygen: that's what rust is all about. So on none of the new planets forming in any solar system can higher life-forms develop until enough oxygen has been pumped into its atmosphere to oxidise most of its free iron. This, not surprisingly, is exactly what the prokaryotes did during their first 2.0 billion years on Earth. But it had to be a two-part process.

The proto-Earth would be cooling the whole time, so let's say full cooling takes roughly 1.0 billion years. So the extremophiles would be the first batch of prokaryotes inserted because they could survive it. Then, after a billion years or so, the terraformers return and drop off the rest of the prokaryotes, the ones that can live in milder conditions. Also, they have to keep returning on a regular basis because each planet would cool at a different rate due to their different sizes and different physical compositions.

However many "check-up" trips are required, by 2.0 billion years after their first seeding of the new solar system the terraformers realise the third planet from the sun is the only one thriving. They are not surprised, having learned that a "zone of life" exists around *all* suns, regardless of size or type. Now that this sun has taken its optimum shape, they could have predicted which planet or planets would thrive. In this system, the third is doing well but the fourth one is struggling. It has its prokaryotes and it has water, but its abundance of iron (the "red" planet) will require longer to neutralise than such a small planet with a non-reactive core will require to cool off, so it will lose its atmosphere to dissipation into space before a balance can be achieved. The fourth planet will become a wasteland.

The terraformers carry out the next phase of planet-building on the thriving third by depositing larger, more complex, more biologically reactive eukaryotes to accelerate the oxidation process. Eukaryotes are far more fragile than prokaryotes, so they can't be put onto a forming planet until it is sufficiently cooled to have abundant land and water. But once in place and established, their large size (relative to prokaryotes) can metabolise much more oxygen per unit. Together, the fully proliferated prokaryotes and eukaryotes can spew out enough oxygen to oxidise every bit of free iron *on* the Earth's crust and *in* its seas, and before long be lacing the atmosphere with it.

Sure enough, when the terraformers return in another 1.4 billion years they find Earth doing well, but the situation on Mars is unimproved: rust as far as the eye can see. (Mars is likely to have at least prokaryotic life, because there wouldn't have been enough oxygen in the surface water it once had - or in the permafrost it still has - to turn its entire surface into iron oxide.) Earth, however, is doing fine. Most of its free iron is locked up as rust, and oxygen levels in the atmosphere are measurably increasing. It's still too soon to think about depositing highly complex life, but that day is not far off now, measurable in tens of millions of years rather than in hundreds of millions. For the moment, Earth is ready for its first load of multicellular life, and so it is deposited: the Ediacaran Fauna.

Though scientists today have no clear understanding of what the Ediacarans were or what their purpose may have been (because they don't exist today), it seems safe to assume they were even more prolific creators of oxygen than the eukaryotes.

If, indeed, terraformers are behind the development of life on Earth, nothing else makes sense. If, on the other hand, everything that happened here did so by nothing but blind chance and coincidence, it was the most amazing string of luck imaginable. Everything happened exactly when it needed to happen, exactly where it needed to happen, exactly how it needed to happen.

If that's not an outright miracle, I don't know what is.

MAKING BETTER SENSE

Assuming terraformers were/are responsible for seeding and developing life on Earth, we can further assume that by 550 million years ago at least the early oceans were sufficiently oxygenated to support genuinely complex life. That was delivered *en masse* during the otherwise inexplicable Cambrian Explosion, after which followed the whole panoply of "higher" forms of life on Earth as we have come to know it. (The whys and wherefores of that process are, regrettably, beyond the scope of this essay, but there are answers that have as much apparent sense behind them as what has been outlined.)

During those 550 million years, five major and several minor extinction events occurred, after each of which a few million years would pass while the Earth stabilised with environments modified in some way by the catastrophes. Some pre-event life-forms would persist into the new environments, to be joined by new ark-loads delivered by the terraformers, who would analyse the situation on the healing planet and deliver species they knew would survive in the new environments and establish a balance with the life-forms already there (the Interventionist version of punctuated equilibrium).

We've already seen the difficulties Darwinists have with trying to explain the flow of life on Earth presented in the fossil record. That record *can* be explained by the currently accepted Darwinian paradigm, but the veneer of "scholarship" overlaying it is little different from the divine fiat of Creationists. And it can be explained by Intelligent Designers, who claim anything so bewilderingly complex couldn't possibly have been arrayed without the guidance of some superior, unifying intelligence (which they stop short of calling "God", because otherwise they are merely Creationists without cant).

Considering all of the above, we Interventionists believe the terraformer scenario explains the fossil record of life on Earth with more creativity, more accuracy and more logic than the others, and in the fullness of time will have a far greater probability of being proved correct. We don't bother trying to establish or even discuss who the terraformers are, or how they came to be, because both are irrelevant and unknowable until they choose to explain it to us. Besides, speculating about *their* origin detracts from the far more germane issue of trying to establish that our explanation of life's origin makes better sense than any other.

We will continue to be ignored by mainstream media simply because the idea of intelligent life existing outside Earth is so frightening to the majority of those bound to it. Among many reasons for fear, the primary one might be our unfortunate habit of filtering everything beyond our immediate reality through our own perceptions. Thus, we attribute to others the same traits and characteristics we possess. Another bad habit appears when we discover new technology. Invariably our first thought is: "How can we use this to kill more of our enemies?" Collectively, we all have enemies we want to eliminate to be done with the problem they present. Like it or not, this is a dominant aspect of human nature.

Because we so consistently project onto others the darkest facets of our nature, we automatically assume - despite ET and Alf and other lovable depictions in our culture - that *real* aliens will want to harm us. Consequently, we avoid facing the possibility of their existence in every way we can. (Here I can mention the obstinate resistance I have personally found to serious consideration of the Starchild skull, which by all rights should have been eagerly and thoroughly examined three years ago.)

So Interventionism is ignored because it scrapes too close to UFOs, crop circles, alien abductions and every other subject that indicates we humans may, in the end, be infinitesimally insignificant in the grand scheme of life in the universe. There is much more to say about it, of course, especially as it relates to human origins, but that has to wait until the second installment of this essay.

For now, let the last word be that the last word on origins - of life and of humans - is a long, long way from being written.

But when it is, I strongly suspect it will be *Intervention*.

PART TWO

HUMAN ORIGINS: CAN WE HANDLE THE TRUTH?

In Part One of this essay, I explained the Interventionist perspective regarding the origin of life on Earth. I showed how the great preponderance of evidence indicates life *came* here and did not *develop* here, as we have been brainwashed to believe by generations of scientists struggling to keep the creation myths of religion out of classrooms. Personally, I applaud and support all efforts to keep the most specious aspects of Creationism safely bottled up in houses of worship, where they belong. However, I have even more disdain for scientists who allow themselves to be crushed to cowardly pulp by nothing more debilitating than "peer pressure". Because both groups are so driven by their collective fears and dogma, neither has a working grip on reality. That becomes increasingly clear as research continues, which I believe was made evident in Part One. Now let's try to do the same in Part Two, on human origins.

If anything riles Creationists and Darwinists alike, it's the suggestion they might be wrong about how we humans have come to dominate our planet so thoroughly. Both sides can tolerate substantial criticisms regarding the wide array of subjects under their purviews, including the kind of critique I gave the origins of life in Part One.

However, they have *no* toleration for challenges to their shared hegemony over the beginnings of us all. Dare that and you'll find yourself in a serious fight. Thus, those of us who support the Interventionist interpretation come under attack from both sides, not to mention the other clique at the party, the educated subgroup of Creationists known as Intelligent Designers (a brilliant choice of name that enforces their bottom-line concept of a "grand designer", while simultaneously implying they are smarter than anyone who would oppose them).

All sides seem to agree that humans are "special". Creationists and Intelligent Designers consider it virtually self-evident that humans originated by some kind of divine fiat. Creationists believe the instigator is a universal "godhead" figure, which IDers water down to a more palatable "entity or system" capable of generating order out of chaos, life out of the inanimate. Even Darwinists will concede that many of our physical, emotional and intellectual traits set us far apart from the primate ancestors they believe preceded us in the biological process of evolution. However, despite our high degree of "specialness", Darwinists fervently promote the dogma that even the most fanciful distinctions separating us from our supposed ancestors can be explained entirely by "natural means".

As with the early life-forms discussed in Part One, there's nothing natural about it.

THE EARLIEST PRIMATES

Darwinists believe the human saga begins with mouse-sized mammals called *insectivores* (similar to modern tree shrews) that scurried around under the feet of large dinosaurs, trying to avoid becoming food for smaller species. Then comes the Cretaceous extinction event of 65 million years ago that took out the dinosaurs and paved the way for those tiny insectivores to evolve over the next few million years into the earliest primates, the prosimians (literally *pre-simians*, pre-monkeys) of the early Palaeocene epoch, which lasts until 55 million years ago.

As with nearly all such aspects of Darwinist dogma, this is pure speculation. There is, in fact, *no* clear indication of a transitional insectivore-to-prosimian species at any point in the process. If any such transitional species had *ever* been found, then countless more would be known and I wouldn't be writing this essay. Darwinian evolution would be proved beyond doubt, and that would be the end of it.

To read the fossil record literally is to discover the legitimacy of *punctuated equilibrium* (discussed in Part One) as a plausible explanation. "Punk eek", as detractors call it, points out that in the fossil record life-forms *do* seem simply to appear on Earth, most often after extinction events but not always. Both the supposed proto-primates and flowering plants appear during the period *preceding* the Cretaceous extinction. They come when they come, so the relatively sudden post-extinction appearance of the earliest primates, the prosimians (lemurs, lorises, tarsiers), is one of many sudden manifestations.

In terms of human origins, it begs this question: did proto-primates actually *evolve* into prosimians, into monkeys, into apes, into humans? Or did prosimians *appear*, monkeys *appear*, apes *appear*, and humans *appear*? Or, in our "special" case, were we created?

However it happened, there is a pattern. The earliest prosimians are found in the fossil record after the Mesozoic/Cenozoic boundary at 65 million years ago. It is *assumed* their ancestors will someday be found as one of countless "missing links" needed to make an airtight case for Darwinian evolution. Prosimians dominate through the Palaeocene and the Eocene, lasting from 65 to 35 million years ago. (There won't be a test on terms or dates, so don't worry about memorising them; just try to keep the time-flow in mind.) At 35 million years ago, the Oligocene epoch begins and the first monkeys come with it.

Again, Science *assumes* that monkeys evolved from prosimians, even though evidence of that transition is nowhere in sight. In fact, there is strong evidence pointing in the other direction, toward the dreaded *stasis* of punctuated equilibrium. The lemurs, lorises and tarsiers of today are essentially just as they were 50 million years ago. Some species have gone extinct while others have modified into new forms, but lemurs and lorises still have wet noses and tarsiers still have dry, which seems always to have been the case. That's why tarsiers are assumed to be responsible for spinning off monkeys and all the rest.

Monkeys start appearing at 35 million years ago, looking vastly different from prosimians. There are certain physiological links, to be sure, such as grasping hands and feet to permit easy movement through trees. However, prosimians cling and jump to move around, while monkeys favour brachiating - swinging along by their arms. Also, prosimians live far more by their sense of smell than do monkeys. This list goes on.

The reason they're linked in an evolutionary flowchart is because they seem close enough in enough ways to make the linkage stick. Simple as that. Science focuses on the similarities and tries hard to ignore their gaping discrepancies, assuming - as always - that there is plenty of time for evolution to do its magic and generate those inexplicable differences.

For the next 10 million years the larger, stronger, more "advanced" monkeys compete with prosimians for arboreal resources, quickly gaining the upper hand over their "ancestors" and driving several of them to extinction.

Then, at around 25 million years ago, the Miocene epoch brings the first apes into the fossil record, as suddenly and inexplicably as all other primates appear. Again, Science *insists* they evolved from monkeys, but the evidence to support that claim is as specious as the prosimian-monkey link. The transitional bones needed to support it are simply *not* in the fossil record.

If this isn't a distinct pattern of punctuated equilibrium, then what is?

THE PUZZLING MIOCENE

In terms of primate evolution, the Miocene makes little sense. By 25 million years ago, when it begins, prosimians have been around for about 30 million years and monkeys for 10 million years. Yet in the Miocene's ample fossil record, prosimians and monkeys are rare, while the new arrivals, the apes, are all over the place.

The Miocene epoch stretches from 25 million to 5.0 million years ago. (These are approximations quoted differently in various sources; I round off to the easiest numbers to keep track of.) During those 20 million years, the apes flourish. They produce two-dozen different genera (types), and many have more than one species within the genus. Those apes come in the same range of sizes they exhibit today, from smallish gibbon-like creatures, to mid-range chimp-sized ones, to large gorilla-sized ones, to super-sized *Gigantopithecus*, known only by many teeth and a few mandibles (jawbones) from India and China.

That's another interesting thing about Miocene apes: their fossils are found literally everywhere in the Old World - Africa, Europe, Asia. Most of them are known by the durable teeth and jaws that define *Gigantopithecus*, while many others supply enough post-cranial (below the head) bones to grant a reasonably clear image of them. They present an interesting mix of anatomical features. Actually, "confusing" is more like it. They are clearly different from monkeys in that they have no tails, just like modern apes. However, their arms tend to be more like monkey arms - the same length as their legs. Modern ape arms are significantly longer than their legs so they can "walk" comfortably on their front knuckles. More than any other reason, this is why we hear so little from anthropologists about Miocene apes. Their arms don't make sense as the forelimbs of an ancestral quadruped. Miocene arms fit better with something else.

This is not to say, of course, that *no* ape arms in the Miocene fossil record are longer than legs. That's nowhere near to being determined because many species - like *Gigantopithecus* - have yet to provide their arm bones. However, since we *do* have some tailless, ape-like bodies with monkey-like arms and hands, we have to consider how such a hybrid would move around. Swing through trees by its arms, like a monkey? Not likely. Monkey arms are designed to carry a monkey's slight body. An ape's body needs to be brachiated and leveraged by an ape's much longer, stouter, stronger arms. So how about *walking*?

From a physiological standpoint, an ape-like body with monkey-like arms and hands does not move as easily or comfortably as a quadruped (down on all fours). It simply can't happen. In fact, there's really only one posture that lends itself to the carriage of such a monkey-ape hybrid, and that's *upright*. Go to a zoo and watch how much easier monkeys - tails and all - stand upright compared to apes. Any monkey can move with grace on its hind legs. In comparison, apes are blundering, top-heavy oafs. Thus, it seems likely that at least some of the hybrid monkey-apes of the Miocene probably had to carry themselves upright, in opposition to the other apes of the era bearing the longer, thicker arms of gibbons, orang-utans, chimpanzees and gorillas. Remember, we're talking about two dozen genera and around 50 species.

WALKING THE WALK

Walking is critical to an understanding of human origins because Darwinists feel it is *the* factor that set our ancestors on the road to becoming us. The theory is that around 5.0 to 10 million years ago, when the heavy forests blanketing Africa began shrinking, some forest-dwelling quadrupedal Miocene apes still living then (there had been the inevitable extinctions and speciations during the preceding 15 to 20 million years) began to forage on the newly forming savannas. Though terribly ill-equipped to undertake such a journey (more about that later), several ape species supposedly took the risk by learning to stand upright to see out over the savanna grasses to scout for predators. Then - after millennia of holding that position for extended periods - they adopted constant upright posture. In doing so, one of those daring, unknown species took the *real* "giant step for mankind".

No one can yet say which of the early upright-walking "pre-humans" went on to become us, because the physiological gaps between us and them are simply enormous. In fact, physically, the only significant thing we have in common with those early ancestors is upright posture. But even that reveals noticeable divergence.

Incredibly, we have the walking trail of at least two early pre-humans at 3.5 million years ago. Found in Laetoli, Tanzania, these tracks were laid down on a volcanic ash fall that was then covered by another ash fall and sealed until their discovery by Mary Leakey's team in 1978. Photos of that trail are common and can be accessed in any basic anthropology textbook or on the Internet. What is not commonly portrayed, however, is that detailed analysis of the pressure points along the surface of those prints indicates something that would be expected: they didn't walk like us. After all, 3.5 million years is a long time, and from a Darwinist standpoint it's logical to assume extensive evolution would occur. But whether it was evolution or not, our methods of locomotion are uniquely different.

Humans have a distinctive carriage that starts with a heel strike necessitated by our ankles placed well behind the midpoint of our feet. After the heel strike, our forward momentum is swung to the left or right, out to the edges of our feet to avoid our arches (in normal feet, of course). Once past the arch, there's a sharp swing of the momentum through the ball of the foot from outside all the way to the inside, where momentum is gathered and regenerated in the powerful thrust of the big toe, with the four small toes drawing themselves up to act as balancers. (Watch your own bare feet when you take a step and you'll see those final "thrust-off" stages in action.)

The pre-humans at Laetoli walked with marked differences. Instead of having a heavy heel-strike leading the way, their ankle was positioned at the centre balance point of the foot, allowing it to come down virtually flat with an almost equal distribution of weight and momentum between the heel and the ball area. Instead of a crazy momentum swing out and around the arch, their arches were much smaller and the line of momentum travelled nearly straight along the midline of the entire foot. That made for a much more stable platform for planting the foot and toeing off into the next step, which was done by generating thrust with the entire ball area rather than with just the big toe. When you get right down to it, the Laetoli stride was a superior technique to the one we utilise now.

Slow-motion studies of humans walking show that we do virtually everything "wrong". Our "heel-strike, toe-off" causes a discombobulation that courses up our entire body. We are forced to lock our knees to handle the torque as our momentum swings out and around our arches. Because of that suspended moment of torque absorption, we basically have to fall forward with each step, which is absorbed by our hip joints. Meanwhile, balance is assisted by swinging our arms. Because of those factors, we don't walk with anything approaching optimum efficiency, and the stresses created in us work, over time, to deteriorate our joints and eventually cripple us. In short, we could use a re-design.

What we actually need to do is to walk more like the pre-humans at Laetoli. In order to secure that heel-and-toe plant with each step, we'd have to modify our stride so our knees weren't locked and we weren't throwing ourselves forward through our hip joints. We'd have to keep our knees in a state of continual flexion, however slight, absorbing all the stress of walking in our thighs and buttocks, which both are designed to accommodate. This would provide us with a "gliding" kind of stride that might look unusual (it would resemble the classic Groucho Marx bent-kneed comedic walk), but would actually be much less stressful, much less tiring and incredibly more efficient physiologically.

Based on the evidence of the Laetoli tracks, this is exactly how they walked.

WHAT'S WRONG WITH THIS PICTURE?

When Darwinists present reconstructions of so-called "pre-humans", invariably they look nothing *like* humans.

Lucy and her *Australopithecus* relatives were little more than upright-walking chimpanzees. The robust australopithecines were bipedal gorillas. The genus *Homo* (*habilis*, *erectus*, Neanderthals and other debatable species) was a distinct upgrade, but still nowhere near the ballpark of humanity. Only when the Cro-Magnons appear, as suddenly and inexplicably as everything else, at around 120,000 years ago in the fossil record, do we see beings that are unmistakably human.

The Laetoli walkers lived 3.5 million years ago. Lucy lived around 3.2 million years ago. Recent discoveries show signs of pushing bipedal locomotion back as far as 6.0 million years ago. So let's assume for the sake of discussion that *some* primates were upright at no less than 4.0 million years ago.

Thus, from approximately 4.0 million years ago all the way to the appearance of Cro-Magnons some time before 120,000 years ago (95% of the journey), all pre-human fossils reveal distinctly *non*-human characteristics. They have thick, robust bones - much thicker and more robust than ours. Such thick bones are necessary to support the stress generated by extraordinarily powerful muscles, far more powerful than ours. Their arms are longer than ours, especially from shoulder to elbow. Their arms are also roughly the same length as their legs, *à la* Miocene apes. And in every aspect that can be quantified - every one! - their skulls are much more ape-like than human-like. Those differences hold from australopithecine bones to the bones of Neanderthals - which means that something quite dramatic happened to produce the Cro-Magnons, and it wasn't the result of an extinction event. It was something else.

The chasm between Cro-Magnons (us) and everything else that comes before them is so incredibly wide and deep that there is no way legitimately to connect the two, apart from linking their bipedal locomotion. All of the so-called "pre-humans" are much more like upright-walking chimps or upright-walking gorillas than they are incipient humans. Darwinists argue that this is why they are called *pre*-humans, because they are so clearly *not* human.

But another interpretation can be put on the fossil record - one that fairly and impartially judges the facts as they exist, without the "spin" required by Darwinist dogma. That spin says that the gaping physiological chasm between Neanderthals and Cro-Magnons can be plausibly explained with yet another "missing link".

LOOKING BACK TO SEE AHEAD

Darwinists use the missing link to negate the fact that Cro-Magnons appear out of nowhere, looking nothing like anything that has come before. What they fail to mention is that *dozens* of such links would be needed to show *any* kind of plausible transition from any pre-human to Cro-Magnons. It clearly didn't happen - and since they're experts about such things, they *know* it didn't happen. However, to acknowledge *that* would play right into the desperate hands of Creationists and Intelligent Designers, not to mention give strong support to Interventionists like me. They face a very big rock or a very hard place.

Let's accept for the moment that in Darwinian terms there is no way to account for the sudden appearance of Cro-Magnons (humans) on planet Earth. If that is true, then what about the so-called "pre-humans"? What are they the ancestors of? Their bones litter the fossil record looking very unlike humans, yet they clearly walk upright for at least 4.0 million years, and new finds threaten to push that back to 6.0 million years. Even more likely is that among the 50 or more species of Miocene apes, at least a few are walking upright as far back as 10 to 15 million years ago. If we accept that likelihood, we finally make sense of the deep past while beginning for the first time to see ourselves clearly.

We can be sure that at least four of the 50 Miocene apes were on their way to becoming modern quadrupeds, because their descendants live among us today. Equally certain is that others of those 50 walked out of the Miocene on two legs. Technically these are called *hominoids*, which are human-*like* beings that are clearly *not* human. In fact, every bipedal fossil preceding Cro-Magnon is considered a hominoid - a term that sounds distinctly outside the human lineage. So Darwinists have replaced it in common usage with the much less specific "pre-human", which not so subtly brainwashes us all into believing there is no doubt about that connection. And that brainwashing works.

We are further brainwashed to believe there are no bipedal apes alive in the world today, despite hundreds of sightings and/or encounters with such bipedal apes every year on every continent except Antarctica. Darwinists brainwash us to ignore such reports by showering them with ridicule. They call such creatures "impossible", and hope the weight of their credentials can hold reality at bay long enough for them to figure out what to do about the public relations catastrophe they will face when the first hominoid is brought onto the world stage - dead or alive. That will be the darkest day in Darwinist history, because their long charade will be officially over. The truth will finally be undeniable. Bigfoot, the Abominable Snowman and several relatives are absolutely real.

IF THE SHOE FITS

I'm not going to waste time and space here going over the mountain of evidence that is available in support of hominoid reality. I cover it extensively in the third part of my book, *Everything You Know Is Wrong*, and there are many other books that cover one or more aspects of the subject. If you care to inform yourself about the reality of hominoids, you won't have any trouble doing so. And the evidence is solid enough to hold up in any court in the world, except the court of public opinion manipulated by terrified Darwinists. However, I will go over a few points that bear directly on the question of human origins.

Let's grant a fairly obvious assumption: that the thousands of ordinary people who have described hominoid sightings and encounters over the past few hundred years (yes, they go back that far in the literature) were in fact seeing living creatures rather than Miocene ghosts. And no matter where on Earth witnesses come from, no matter how far from the beaten path of education and/or modern communications, they describe what they see with amazing consistency. To hear witnesses tell it, the same kinds of creatures exist in every heavily forested or canopied environment on the planet - which is precisely what we would expect if they did indeed stride out of the Miocene epoch on two legs.

Furthermore, what witnesses describe is exactly what we would expect of upright-walking apes. They are invariably described as having a robust, muscular body covered with hair, atop which sits a head with astonishingly ape-like features. In short, the living hominoids are described as having bodies we would expect to find wrapped around the bones found in the so-called "pre-human" fossil record. In addition, witnesses describe what they see as having longer arms than human arms, hanging down near their knees, which means those arms are approximately the length of their legs. Witnesses also contend that the creatures walk with a "gliding" kind of bent-kneed stride that leaves tracks eerily reminiscent of the tracks left at Laetoli 3.5 million years ago.

Now we come to the crux for Darwinists, Creationists and Intelligent Designers. Evidence supporting the reality of hominoids is overwhelming. Truly. And if they *are* real, it means the "pre-human" fossil record is actually a record of *their* ancestors, not ours. And if that's the case, then humans have no place on the flowchart of life on Earth. And if that's true, then it's equally clear that humans *did not* evolve and could not have evolved here the way Darwinists claim. And if we didn't evolve here, that opens the door to the Interventionist position that *nothing* evolved here: everything was brought or created by sentient off-world beings whom I call *terraformers*, whose means and motivation will remain unknown to us unless and until they see fit to explain themselves. I hope no one is holding their breath.

The point is that the Miocene epoch had the means to produce living hominoids - 50 or more different species (which almost certainly will be shaved down to perhaps a dozen as more complete bodies are found) as far back as 20 million years ago. It produced some with monkey-like arms better suited to an upright walker than a brachiating tree-dweller or knuckle walker.

By the time it ended, 5.0 million years ago, a half-dozen or more bipedal apes were on the Earth, which we know from the ape-like australopithecine and early *Homo* fossils. And we know from Laetoli that they had a walking pattern distinct from humans, which modern witnesses describe as still being the way hominoids walk. In short, they've followed the punctuated equilibrium pattern of long-term *stasis*.

SO WHAT ABOUT HUMANS?

Humans simply do not fit the pattern of primate development on Earth. Notice the word *development* instead of *evolution*. Species that appear here do undergo changes in morphology over time. It's called *micro*evolution, because it describes changes in body parts. Darwinists use the undeniable reality of microevolution to extrapolate the reality of *macro*evolution, which is change at the species-into-more-advanced-species level. That is blatantly *not* evident in the fossil record, especially when it comes to human physiology.

We have shown, I hope, that humans have been shoehorned by Darwinists into having a place in the fossil record that doesn't belong to them but to living hominoids (Bigfoot, etc.). Furthermore, humans have been shoehorned into being primates, when there is little about them - certainly nothing of significance - that fits the classic primate pattern. In fact, if it weren't for the desperate need of Darwinists to keep humans closely linked to the primate line, we would have had our own appellation long ago - and we'll surely have it once the truth is out from the Pandora's box of Darwinist deception.

Relatively speaking, primate bones are much thicker and heavier than human bones. Primate muscles are five to 10 times stronger than ours. (Anyone who's dealt with monkeys knows how amazingly strong they are for their size.) Primate skin is covered with long, thick, visible hair. Ours is largely invisible. Primate hair is thick on the back, thin on the front. Ours is switched the other way around. Primates have large, round eyes capable of seeing at night. Compared to theirs, we have greatly reduced night vision. Primates have small, relatively "simple" brains compared to ours. They lack the ability to modulate sound into speech. Primate sexuality is based on an oestrus cycle in females (though some, like bonobo chimps, have plenty of sex when not in oestrus). In human females, the effects of oestrus are greatly diminished.

This list could go on to cite many more areas of difference, but all of them are overshadowed by the Big Kahuna of primate/human difference: all primates have 48 chromosomes, while humans have "only" 46 chromosomes. Two entire chromosomes represent a heck of a lot of DNA removed from the human genome, yet somehow that removal made us "superior" in countless ways. It doesn't make sense. Nor does the fact that even with two whole chromosomes missing from our genome, we share what is now believed to be 95% of the chimp genome and around 90% of the gorilla genome. How can those numbers be made to reconcile? They can't.

Something is wrong here. Someone has been cooking the genetic books.

THE STUFF OF LIFE

In the wild, plants and animals tend to breed remarkably true to their species. That's why stasis is the dominant characteristic of life on Earth. Species appear and stay essentially the same (apart from the superficial changes of microevolution) until they go extinct for whatever reason (catastrophe, inability to compete for resources effectively, etc.). When "faulty" examples appear, they're nearly always unable to put the fault into their species' collective gene pool. A negative mutation that doesn't kill the individual it appears in is unlikely to be passed along to posterity, despite Darwinist assertions that this is precisely how evolution occurs. All genomes have hard-wired checks and balances *against* significant changes of *any* kind, which is why *stasis* has been the hallmark of all life since beginning here. Aberrant examples are efficiently weeded out, either early in the reproductive process or soon after reproduction (birth). Faulty copies are deleted.

This deletion of faults holds true in the vast majority of species. Most genomes are - and stay - remarkably clear of gene-based defects. All species are susceptible to mistakes in the reproductive process, such as sperm/egg misconnections. In mammals, this produces spontaneous abortions, stillbirths or live-birth defects. However, there are precious few defects that swim in the gene pools of any "wild" or "natural" species. The only places we find significant, species-wide genetic defects are in *domesticated* plants and animals, and in those they can be - and often are - numerous.

Domesticated plants and animals clearly seem to have been genetically created by "outside intervention" at some point in the distant past. (For those interested in learning more about this, I discuss it in considerable detail in NEXUS 9/04.) Domesticated species have so many points of divergence from wild/natural species, it's not realistic to consider them in any kind of relative context. As we've seen above, the same holds true for humans and the primates we supposedly evolved from. They're apples and oranges.

We humans have over 4,000 genetic defects spread throughout our common gene pool. Think about that. No other species comes close. And yet, our mitochondrial DNA proves we have existed as a species for "only" about 200,000 years. Remember the first Cro-Magnon fossils showing up in strata 120,000 years old? That fits well with the origin of a small proto-group at around 200,000 years ago. (There will almost certainly be Cro-Magnon fossils found prior to 120,000 years ago, but it is unlikely they were dispersed widely enough to have left fossils near the 200,000-year mark. Naturally, the very first one *could* have been fossilised, but that's not the way to bet. Fossilisation is quite rare.)

All that being the case, how did over 4,000 genetic defects work their way into the human gene pool, when such genome-wide defects are rare to nonexistent in wild or natural species? (Remember, Darwin himself noticed that humans are very much like domesticated animals in many of our physical and biological traits.) It can only have occurred if the very first members (no more than a handful of breeding pairs) had the entire package of faults within their genome. That's the only way Eskimos and Watusis and all the rest of humanity can express the exact same genetic disorders.

If we descended from apes, as Darwinists insist, then apes should have a very large number of our genetic defects. They do not. If, on the other hand, we've been genetically unique for only 200,000 years, then the only way those defects could be with us is if they were *put into* our gene pool by the genetic manipulation of the founding generation of our species, and the mistakes made in that process were left in place to be handed down to posterity. And, as might be expected, this is also how domesticated plants and animals came to have their own inordinate numbers of genetic defects. It simply couldn't happen any other way.

THE FINAL NAIL

When Einstein was asked in reference to relativity, "How did you do it?", he replied, "I ignored an axiom." This is what everyone must do if we are to get anywhere near the truth about human origins.

Darwinists ask us to believe a theory based on this axiom: "There are good grounds to believe our early ancestors lived in forests. There are equally good grounds to believe our later ancestors lived by hunting game on African savannas. Therefore, we can assume that somehow, some way, we went from living in forests to living on the savannas." The trick, for Darwinists, is in explaining it plausibly.

Savanna theorists ask us to believe that, 5.0 to 10 million years ago, several groups of forest-dwelling Miocene apes were squeezed by environmental pressures to venture out onto the encroaching savannas to begin making their collective living. This means they had to rise from the assumed quadrupedal posture attributed to all Miocene apes to walk and run on two legs, thus giving up the ease and rapidity of moving on all fours. Those early groups had to make their way with unmodified pelvises, inappropriate single-arched spines, absurdly under-muscled thighs and buttocks, and heads stuck on at the wrong angle, and all the while doggedly shuffling along on the sides of long-toed, ill-adapted feet, thereby becoming plodding skin-bags of snack-treats for savanna predators. If any harebrained scheme ever deserved a re-think by its originator(s), this would be the one.

Of course, the *real* re-think needs to be done by Darwinists, because it is glaringly obvious that no forest-bound species of ape could have ventured onto the savanna as a stumbling, bumbling walker and learned to do it better out there among the big cats. If a collective group had been unfit for erect movement on the savanna, they wouldn't have gone. If they *did* go, they couldn't and wouldn't stay. Even primates are smarter than that. And understand, there are primates that *did* make the move onto the savanna, albeit always remaining within range of a high-speed scurry into nearby trees. Baboons are the most successful of this small group, all of which have retained quadrupedal locomotion.

In addition to the forest-to-savanna transition, Darwinists face numerous other improbable - if not impossible - differences between humans and terrestrial primates. In addition to bipedalism and the genetic discrepancies already addressed, there are major differences in skin and the adipose tissue (fat) beneath it; in sweat glands, in blood, in tears, in sex organs, in brain size and function, and on and on and on. This is a very long list that can be examined in much fuller detail in the work of a brilliant, determined researcher into human origins, named Elaine Morgan.

Ms Morgan is the chief proponent of what challenged Darwinists derisively call "the Aquatic Ape theory", as if the juxtaposition of those disparate words were enough to dismiss it as an absurd notion. Nothing could be further from the truth. In books like *The Scars of Evolution* (Souvenir Press, London, 1990), she makes a devastating case against the notion that humans evolved from forest-dwelling apes that moved out onto the savannas. She believes humans must have gone through an extended period of development in and around water to generate the bizarre array of physiological oddities we exhibit relative to the primates we supposedly evolved from.

However, despite all her wonderfully creative work, Ms Morgan remains wedded to the Darwinist concept of evolution, which had to play itself out in only the 200,000 years dictated by our mitochondrial DNA.

MAKING SENSE OF THE INSENSIBLE

The pieces of the puzzle are on the table. The answer is there for anyone to see. But rearranging those pieces properly is no easy task, and it is even more difficult to get dogmatists of any stripe to look at the picture in a light different from their own. That has been my purpose in writing these two essays on origins - of life and of humans. They are two of the world's most sensitive areas of scholarship and debate, producing some of the most vitriolic exchanges in all of academia. But vitriol, like might, doesn't make right.

I once knew a baseball player who'd pitched a no-hitter against a seriously inferior team. Upon being criticised for the obvious imbalance between his abilities and those of his opponents, the pitcher shrugged and said, "A no-hitter is a no-hitter, even against Lighthouse for the Blind." And so it is with a mistaken belief. If millions believe a thing, that doesn't make it correct.

I believe that the facts, if fairly evaluated, will over time prove that humans - and indeed, life itself - did *not* originate on Earth, and that *nothing* has macroevolved on Earth. It has all been brought here and left to fend for itself, then replaced when events required the introduction of new forms. No other theory suits the facts nearly as well.

As for humans (the object of this essay), look back to the Miocene epoch, where the earliest traces of our ancestors supposedly originate. Apes dominate. Look at the fossils - the so-called "pre-humans" - from the Pliocene epoch, starting 5.0 million years ago. Other than bipedal walking, all of their physical aspects shout out "ape roots". Look at today's tracks, sightings and encounters with living hominoids, Bigfoot and others. These all-too-real creatures will one day be proved to have a direct link back to the Miocene - which, at a stroke, will eliminate any possibility that humans and apes share any kind of common ancestor.

We humans are not indigenous to planet Earth. We were either put here intact or we developed here, but we did not evolve here. Our genes make clear that we've been cut-and-pasted from other, non-primate, non-Earthly species.

Personally, I believe that the work of Zecharia Sitchin (*The Earth Chronicles*) comes closest to a plausible explanation. But even if some aspects of what he says are wrong, or even if all of it someday is proved to be wrong, that won't change the basic facts that his work - and my own work - address.

Humans are not primates. We do indeed stand apart as a "special" creation, long espoused by theologians and now by certain credentialled scientists. The only question left hanging is, of course: who or what was the creator? I don't think I'll be privileged to learn that in my lifetime. But I'm confident I'm within reach of the next best answer.

I'm confident that we were created by invasive genetic manipulation.

About the Author:

Lloyd Pye, born in 1946 in Louisiana, USA, is a researcher, author, novelist and scriptwriter. His independent studies over more than three decades into all aspects of evolution have convinced him that humans did not evolve on Earth, or at least are the product of extraterrestrial intervention. His book, *Everything You Know Is Wrong - Book One: Human Origins*, can be ordered through website http://www.iUniverse.com or Barnes & Noble at http://www.bn.com.

Director's Report 2002

Dear Friends,

Let me start by thanking you all for your support over the last twelve months. They have not been particularly easy ones, but in many ways this has been the most important year in the history of the CFZ. It has been our tenth anniversary year and we decided to mark it by changing our status. Whilst we had always been a private business, we are now a registered non-profit making trust, and are awaiting the results of our application for charitable status.

After the death of our long-time Hon. Consulting Editor, Professor Bernard Heuvelmans last year, we considered looking for a replacement. However, we decided not to even try. Bernard was unique within cryptozoology and was irreplaceable. We therefore created a new post – that of Hon. Life President – and drew up a shortlist of one. What we would have done if he had refused, I don't know, but at the beginning of this year Colonel John Blashford-Snell did us the signal honour of accepting the post. The Colonel sums up everything that the CFZ wish to project during our second decade of existence. He is the quintessential English explorer and he carries the spirit of "Biggles" into the 21st Century. We hope that we can be inspired by him to do likewise.

On a personal note for a moment. In March my mother died, and from the bottom of my heart I would like to thank all of you who wrote, telephoned and e-mailed messages of sympathy to me and to my family. My mother unwittingly launched me on my career as a cryptozoologist. From early childhood she encouraged my interest in the natural world, and serendipitously when I was seven years old she found me a library book called *Myth or Monster,* which entranced me and lured me away from the paths of orthodox natural history forever.

In April we presented an exhibition at the *Fortean Times* UnConvention in Kensington High Street. The exhibition – to mark our tenth anniversary year – featured some exquisite monster models by Anthony James of Nuneaton, and was a great success. At the Unconvention, long time CFZ member Martin Jenkins - a charity consultant of no mean reputation - offered to help us navigate the legal and procedural minefield that we would have to traverse before attaining charitable status.

In May we promoted the third of our annual Weird Weekends. This was great fun for all involved and featured lectures by Larry Warren *(Rendlesham Forest),* Mike Hallowell *(Dragon cults of north eastern England),* Gail-Nina Anderson *(The evolution of the vampire within art history),* Chris Moiser *(Victorian Travelling Menageries),* Nigel Wright *(oddities of UFOlogy),* John Tindsley *(Dr John Dee),* Malcolm Robinson *(Poltergeist investigations),* Bob Mann *(The Reverend Sabine Baring-Gould),* Ian Simmons *(Images of the Vampire)* and others. We raised enough money to purchase an inflatable dinghy and a laptop computer for the CFZ. I would like to take this opportunity to thank George Bishop from the CCCS and Nichola Sullings (our newly appointed fund-raising officer), for their sterling efforts. Without these two it is safe to say that *Weird Weekend III* would never have gone ahead.

At the end of May, Richard and I travelled to Lancashire to begin our investigation into the "monster" of Martin Mere. We spent a week there and returned at the end of July with the full investigation team and a battery of electronic equipment. Under the kind auspices of Pat Wisniewski we spent nearly a week there and made positive contact with the creature both visually and on sonar.

It turned out to be a Wels catfish – an eastern European species that can reach a size of 16 feet although our specimen was only about half this length. As a result of our monster-hunting activities we were thrown into a whirlwind of TV, newspaper and radio appearances which culminated in Richard and me appearing live on GMTV during a rainstorm which would have seemed familiar to Noah! We would particularly like to thank Pat and Louise and Tim and Lynda Matthews for their help and encouragement.

In July we relaunched our journal *Animals & Men* with a bold new graphic image by Mark North. During the year, Mark (together with Graham) also revamped the image of the website. In August we purchased our own domain name of www.cfz.org.uk through the kind auspices of the CFZ Computer Consultant (vulgarly referred to as a "Software Jockey"), Andy Billings.

Also in August, Richard and I travelled to Cleveland to investigate the bizarre case of the Loftus Wallaby Slasher.

We would like to thank Making Time TV, and the Cleveland Constabulary (particularly PC O'Hara and WPC Dicks) for their help. However, the investigation could not have been successfully concluded without the help of vet Simon Beck, and our old friend and Co. Durham Representative, David Curtis. We would like to thank David and his long-suffering wife

Joanne for welcoming us into their house for several days R&R after the investigation was concluded.

In September we issued my book *The Monster of the Mere* that covered our antics and adventures in Lancashire earlier in the year. We also published issue 28 of *Animals & Men*. In October Richard and I attended the Twilight Worlds Hallowe'en celebrations for the third year running and gave brief talks on our latest adventures. In November I lectured at the Totnes Museum Society before the CFZ battened down hatches for the winter months.

As Director, I am inordinately proud of the CFZ, and I feel humbled to be in charge of an organisation that includes amongst its faculty the cream of Fortean Zoological researchers around the world. As we draw near to the end of one year we look forward to the next. In 2003 we have the long awaited expedition to The Gambia pencilled in for the summer, as well as expeditions to the United States and Greece. We would like to thank everyone who has helped us during the year, and especially our clients whose orders for printing work and media consultancy have funded much of our activities. Please continue to support us over the next twelve months…

Fondest regards,

Jonathan Downes

THE CENTRE FOR FORTEAN ZOOLOGY

So, what is the Centre for Fortean Zoology?

We are a non profit-making organisation founded in 1992 with the aim of being a clearing house for information, and coordinating research into mystery animals around the world. We also study out of place animals, rare and aberrant animal behaviour, and Zooform Phenomena; little-understood "things" that appear to be animals, but which are in fact nothing of the sort, and not even alive (at least in the way we understand the term).

Why should I join the Centre for Fortean Zoology?

Not only are we the biggest organisation of our type in the world, but - or so we like to think - we are the best. We are certainly the only truly global Cryptozoological research organisation, and we carry out our investigations using a strictly scientific set of guidelines. We are expanding all the time and looking to recruit new members to help us in our research into mysterious animals and strange creatures across the globe. Why should you join us? Because, if you are genuinely interested in trying to solve the last great mysteries of Mother Nature, there is nobody better than us with whom to do it.

What do I get if I join the Centre for Fortean Zoology?

For £12 a year, you get a four-issue subscription to our journal *Animals & Men*. Each issue contains 60 pages packed with news, articles, letters, research papers, field reports, and even a gossip column! The magazine is A5 in format with a full colour cover. You also have access to one of the world's largest collections of resource material dealing with cryptozoology and allied disciplines, and people from the CFZ membership regularly take part in fieldwork and expeditions around the world.

How is the Centre for Fortean Zoology organized?

The CFZ is managed by a three-man board of trustees, with a non-profit making trust registered with HM Government Stamp Office. The board of trustees is supported by a Permanent Directorate of full and part-time staff, and advised by a Consultancy Board of specialists - many of whom who are world-renowned experts in their particular field. We have regional representatives across the UK, the USA, and many other parts of the world, and are affiliated with other organisations whose aims and protocols mirror our own.

I am new to the subject, and although I am interested I have little practical knowledge. I don't want to feel out of my depth. What should I do?

Don't worry. We were *all* beginners once. You'll find that the people at the CFZ are friendly and approachable. We have a thriving forum on the website which is the hub of an ever-growing electronic community. You will soon find your feet. Many members of the CFZ Permanent Directorate started off as ordinary members, and now work full-time chasing monsters around the world.

I have an idea for a project which isn't on your website. What do I do?

Write to us, e-mail us, or telephone us. The list of future projects on the website is not exhaustive. If you have a good idea for an investigation, please tell us. We may well be able to help.

How do I go on an expedition?

We are always looking for volunteers to join us. If you see a project that interests you, do not hesitate to get in touch with us. Under certain circumstances we can help provide funding for your trip. If you look on the future projects section of the website, you can see some of the projects that we have pencilled in for the next few years.

In 2003 and 2004 we sent three-man expeditions to Sumatra looking for Orang-Pendek - a semi-legendary bipedal ape. The same three went to Mongolia in 2005. All three members started off merely subscribers to the CFZ magazine.

Next time it could be you!

Project Kerinci, Sumatra - 2003
In search of the bipedal ape Orang Pendek

How is the Centre for Fortean Zoology funded?

We have no magic sources of income. All our funds come from donations, membership fees, works that we do for TV, radio or magazines, and sales of our publications and merchandise. We are always looking for corporate sponsorship, and other sources of revenue. If you have any ideas for fund-raising please let us know. However, unlike other cryptozoological organisations in the past, we do not live in an intellectual ivory tower. We are not afraid to get our hands dirty, and furthermore we are not one of those organisations where the membership have to raise money so that a privileged few can go on expensive foreign trips. Our research teams both in the UK and abroad, consist of a mixture of experienced and inexperienced personnel. We are truly a community, and work on the premise that the benefits of CFZ membership are open to all.

What do you do with the data you gather from your investigations and expeditions?

Reports of our investigations are published on our website as soon as they are available. Preliminary reports are posted within days of the project finishing.

Each year we publish a 200 page yearbook containing research papers and expedition reports too long to be printed in the journal. We freely circulate our information to anybody who asks for it.

Is the CFZ community purely an electronic one?

No. Each year since 2000 we have held our annual convention - the *Weird Weekend* - in Exeter. It is three days of lectures, workshops, and excursions. But most importantly it is a chance for members of the CFZ to meet each other, and to talk with the members of the permanent directorate in a relaxed and informal setting and preferably with a pint of beer in one hand. Since 2006 - the *Weird Weekend* has been bigger and better and held in the idyllic rural location of Woolsery in North Devon. The 2008 event will be held over the weekend 15-17 August.

Since relocating to North Devon in 2005 we have become ever more closely involved with other community organisations, and we hope that this trend will continue. We also work closely with Police Forces across the UK as consultants for animal mutilation cases, and we intend to forge closer links with the coastguard and other community services. We want to work closely with those who regularly travel into the Bristol Channel, so that if the recent trend of exotic animal visitors to our coastal waters continues, we can be out there as soon as possible.

We are building a Visitor's Centre in rural North Devon. This will not be open to the general public, but will provide a museum, a library and an educational resource for our members (currently over 400) across the globe. We are also planning a youth organisation which will involve children and young people in our activities. We work closely with *Tropiquaria* - a small zoo in north Somerset, and have several exciting conservation projects planned.

Apart from having been the only Fortean Zoological organisation in the world to have consistently published material on all aspects of the subject for over a decade, we have achieved the following concrete results:

- Disproved the myth relating to the headless so-called sea-serpent carcass of Durgan beach in Cornwall 1975
- Disproved the story of the 1988 puma skull of Lustleigh Cleave
- Carried out the only in-depth research ever into the mythos of the Cornish Owlman
- Made the first records of a tropical species of lamprey
- Made the first records of a luminous cave gnat larva in Thailand.
- Discovered a possible new species of British mammal - the beech marten.
- In 1994-6 carried out the first archival fortean zoological survey of Hong Kong.
- In the year 2000, CFZ theories where confirmed when an entirely new species of lizard was found resident in Britain.
- Identified the monster of Martin Mere in Lancashire as a giant wels catfish
- Expanded the known range of Armitage's skink in the Gambia by 80%
- Obtained photographic evidence of the remains of Europe's largest known pike
- Carried out the first ever in-depth study of the *ninki-nanka*
- Carried out the first attempt to breed Puerto Rican cave snails in captivity
- Were the first European explorers to visit the `lost valley` in Sumatra
- Published the first ever evidence for a new tribe of pygmies in Guyana
- Published the first evidence for a new species of caiman in Guyana

EXPEDITIONS & INVESTIGATIONS TO DATE INCLUDE:

- 1998 Puerto Rico, Florida, Mexico *(Chupacabras)*
- 1999 Nevada *(Bigfoot)*
- 2000 Thailand *(Giant snakes called nagas)*
- 2002 Martin Mere *(Giant catfish)*
- 2002 Cleveland *(Wallaby mutilation)*
- 2003 Bolam Lake *(BHM Reports)*
- 2003 Sumatra *(Orang Pendek)*
- 2003 Texas *(Bigfoot; giant snapping turtles)*
- 2004 Sumatra *(Orang Pendek; cigau, a sabre-toothed cat)*
- 2004 Illinois *(Black panthers; cicada swarm)*
- 2004 Texas *(Mystery blue dog)*
- 2004 Puerto Rico *(Chupacabras; carnivorous cave snails)*
- 2005 Belize *(Affiliate expedition for hairy dwarfs)*
- 2005 Mongolia *(Allghoi Khorkhoi aka Mongolian death worm)*
- 2006 Gambia *(Gambo - Gambian sea monster, Ninki Nanka and Armitage s skink*
- 2006 Llangorse Lake *(Giant pike, giant eels)*
- 2006 Windermere *(Giant eels)*
- 2007 Coniston Water *(Giant eels)*
- 2007 Guyana *(Giant anaconda, didi, water tiger)*

To apply for a <u>FREE</u> information pack about the organisation and details of how to join, plus information on current and future projects, expeditions and events.

Send a stamped and addressed envelope to:

**THE CENTRE FOR FORTEAN ZOOLOGY
MYRTLE COTTAGE, WOOLSERY,
BIDEFORD, NORTH DEVON
EX39 5QR.**

or alternatively visit our website at:
www.cfz.org.uk

Other books available from
CFZ PRESS

THE OWLMAN AND OTHERS - 30th Anniversary Edition
Jonathan Downes - ISBN 978-1-905723-02-7

£14.99

EASTER 1976 - Two young girls playing in the churchyard of Mawnan Old Church in southern Cornwall were frightened by what they described as a "nasty bird-man". A series of sightings that has continued to the present day. These grotesque and frightening episodes have fascinated researchers for three decades now, and one man has spent years collecting all the available evidence into a book. To mark the 30th anniversary of these sightings, Jonathan Downes has published a special edition of his book.

DRAGONS - More than a myth?
Richard Freeman - ISBN 0-9512872-9-X

£14.99

First scientific look at dragons since 1884. It looks at dragon legends worldwide, and examines modern sightings of dragon-like creatures, as well as some of the more esoteric theories surrounding dragonkind.

Dragons are discussed from a folkloric, historical and cryptozoological perspective, and Richard Freeman concludes that: "When your parents told you that dragons don't exist - they lied!"

MONSTER HUNTER
Jonathan Downes - ISBN 0-9512872-7-3

£14.99

Jonathan Downes' long-awaited autobiography, *Monster Hunter*...

Written with refreshing candour, it is the extraordinary story of an extraordinary life, in which the author crosses paths with wizards, rock stars, terrorists, and a bewildering array of mythical and not so mythical monsters, and still just about manages to emerge with his sanity intact.......

MONSTER OF THE MERE
Jonathan Downes - ISBN 0-9512872-2-2

£12.50

It all starts on Valentine's Day 2002 when a Lancashire newspaper announces that "Something" has been attacking swans at a nature reserve in Lancashire. Eyewitnesses have reported that a giant unknown creature has been dragging fully grown swans beneath the water at Martin Mere. An intrepid team from the Exeter based Centre for Fortean Zoology, led by the author, make two trips – each of a week – to the lake and its surrounding marshlands. During their investigations they uncover a thrilling and complex web of historical fact and fancy, quasi Fortean occurrences, strange animals and even human sacrifice.

**CFZ PRESS, MYRTLE COTTAGE,
WOOLFARDISWORTHY BIDEFORD,
NORTH DEVON, EX39 5QR
www.cfz.org.uk**

Other books available from
CFZ PRESS

ONLY FOOLS AND GOATSUCKERS
Jonathan Downes - ISBN 0-9512872-3-0

£12.50

In January and February 1998 Jonathan Downes and Graham Inglis of the Centre for Fortean Zoology spent three and a half weeks in Puerto Rico, Mexico and Florida, accompanied by a film crew from UK Channel 4 TV. Their aim was to make a documentary about the terrifying chupacabra - a vampiric creature that exists somewhere in the grey area between folklore and reality. This remarkable book tells the gripping, sometimes scary, and often hilariously funny story of how the boys from the CFZ did their best to subvert the medium of contemporary TV documentary making and actually do their job.

WHILE THE CAT'S AWAY
Chris Moiser - ISBN: 0-9512872-1-4

£7.99

Over the past thirty years or so there have been numerous sightings of large exotic cats, including black leopards, pumas and lynx, in the South West of England. Former Rhodesian soldier Sam McCall moved to North Devon and became a farmer and pub owner when Rhodesia became Zimbabwe in 1980. Over the years despite many of his pub regulars having seen the "Beast of Exmoor" Sam wasn't at all sure that it existed. Then a series of happenings made him change his mind. Chris Moiser—a zoologist—is well known for his research into the mystery cats of the westcountry. This is his first novel.

CFZ EXPEDITION REPORT 2006 - GAMBIA
ISBN 1905723032

£12.50

In July 2006, The J.T.Downes memorial Gambia Expedition - a six-person team - Chris Moiser, Richard Freeman, Chris Clarke, Oll Lewis, Lisa Dowley and Suzi Marsh went to the Gambia, West Africa. They went in search of a dragon-like creature, known to the natives as `Ninki Nanka`, which has terrorized the tiny African state for generations, and has reportedly killed people as recently as the 1990s. They also went to dig up part of a beach where an amateur naturalist claims to have buried the carcass of a mysterious fifteen foot sea monster named 'Gambo', and they sought to find the Armitage's Skink (*Chalcides armitagei*) - a tiny lizard first described in 1922 and only rediscovered in 1989. Here, for the first time, is their story.... With an forward by Dr. Karl Shuker and introduction by Jonathan Downes.

BIG CATS IN BRITAIN YEARBOOK 2006
Edited by Mark Fraser - ISBN 978-1905723-01-0

£10.00

Big cats are said to roam the British Isles and Ireland even now as you are sitting and reading this. People from all walks of life encounter these mysterious felines on a daily basis in every nook and cranny of these two countries. Most are jet-black, some are white, some are brown, in fact big cats of every description and colour are seen by some unsuspecting person while on his or her daily business. 'Big Cats in Britain' are the largest and most active group in the British Isles and Ireland This is their first book. It contains a run-down of every known big cat sighting in the UK during 2005, together with essays by various luminaries of the British big cat research community which place the phenomenon into scientific, cultural, and historical perspective.

CFZ PRESS, MYRTLE COTTAGE,
WOOLSERY, BIDEFORD,
NORTH DEVON, EX39 5QR
w w w . c f z . o r g . u k

Other books available from
CFZ PRESS

THE SMALLER MYSTERY CARNIVORES OF THE WESTCOUNTRY
Jonathan Downes - ISBN 978-1-905723-05-8

£7.99

Although much has been written in recent years about the mystery big cats which have been reported stalking Westcountry moorlands, little has been written on the subject of the smaller British mystery carnivores. This unique book redresses the balance and examines the current status in the Westcountry of three species thought to be extinct: the Wildcat, the Pine Marten and the Polecat, finding that the truth is far more exciting than the currently held scientific dogma. This book also uncovers evidence suggesting that even more exotic species of small mammal may lurk hitherto unsuspected in the countryside of Devon, Cornwall, Somerset and Dorset.

THE BLACKDOWN MYSTERY
Jonathan Downes - ISBN 978-1-905723-00-3

£7.99

Intrepid members of the CFZ are up to the challenge, and manage to entangle themselves thoroughly in the bizarre trappings of this case. This is the soft underbelly of ufology, rife with unsavoury characters, plenty of drugs and booze." That sums it up quite well, we think. A new edition of the classic 1999 book by legendary fortean author Jonathan Downes. In this remarkable book, Jon weaves a complex tale of conspiracy, anti-conspiracy, quasi-conspiracy and downright lies surrounding an air-crash and alleged UFO incident in Somerset during 1996. However the story is much stranger than that. This excellent and amusing book lifts the lid off much of contemporary forteana and explains far more than it initially promises.

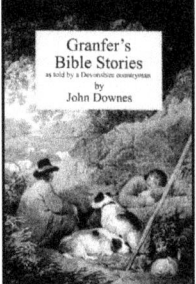

GRANFER'S BIBLE STORIES
John Downes - ISBN 0-9512872-8-1

£7.99

Bible stories in the Devonshire vernacular, each story being told by an old Devon Grandfather - 'Granfer'. These stories are now collected together in a remarkable book presenting selected parts of the Bible as one more-or-less continuous tale in short 'bite sized' stories intended for dipping into or even for bed-time reading. `Granfer` treats the biblical characters as if they were simple country folk living in the next village. Many of the stories are treated with a degree of bucolic humour and kindly irreverence, which not only gives the reader an opportunity to re-evaluate familiar tales in a new light, but do so in both an entertaining and a spiritually uplifting manner.

FRAGRANT HARBOURS DISTANT RIVERS
John Downes - ISBN 0-9512872-5-7

£12.50

Many excellent books have been written about Africa during the second half of the 19th Century, but this one is unique in that it presents the stories of a dozen different people, whose interlinked lives and achievements have as many nuances as any contemporary soap opera. It explains how the events in China and Hong Kong which surrounded the Opium Wars, intimately effected the events in Africa which take up the majority of this book. The author served in the Colonial Service in Nigeria and Hong Kong, during which he found himself following in the footsteps of one of the main characters in this book; Frederick Lugard – the architect of modern Nigeria.

**CFZ PRESS, MYRTLE COTTAGE,
WOOLFARDISWORTHY BIDEFORD,
NORTH DEVON, EX39 5QR
w w w . c f z . o r g . u k**

Other books available from
CFZ PRESS

ANIMALS & MEN - Issues 1 - 5 - In the Beginning
Edited by Jonathan Downes - ISBN 0-9512872-6-5

£12.50

At the beginning of the 21st Century monsters still roam the remote, and sometimes not so remote, corners of our planet. It is our job to search for them. The Centre for Fortean Zoology [CFZ] is the only professional, scientific and full-time organisation in the world dedicated to cryptozoology - the study of unknown animals. Since 1992 the CFZ has carried out an unparalleled programme of research and investigation all over the world. We have carried out expeditions to Sumatra (2003 and 2004), Mongolia (2005), Puerto Rico (1998 and 2004), Mexico (1998), Thailand (2000), Florida (1998), Nevada (1999 and 2003), Texas (2003 and 2004), and Illinois (2004). An introductory essay by Jonathan Downes, notes putting each issue into a historical perspective, and a history of the CFZ.

ANIMALS & MEN - Issues 6 - 10 - The Number of the Beast
Edited by Jonathan Downes - ISBN 978-1-905723-06-5

£12.50

At the beginning of the 21st Century monsters still roam the remote, and sometimes not so remote, corners of our planet. It is our job to search for them. The Centre for Fortean Zoology [CFZ] is the only professional, scientific and full-time organisation in the world dedicated to cryptozoology - the study of unknown animals. Since 1992 the CFZ has carried out an unparalleled programme of research and investigation all over the world. We have carried out expeditions to Sumatra (2003 and 2004), Mongolia (2005), Puerto Rico (1998 and 2004), Mexico (1998), Thailand (2000), Florida (1998), Nevada (1999 and 2003), Texas (2003 and 2004), and Illinois (2004). Preface by Mark North and an introductory essay by Jonathan Downes, notes putting each issue into a historical perspective, and a history of the CFZ.

BIG BIRD! Modern Sightings of Flying Monsters

Ken Gerhard - ISBN 978-1-905723-08-9

£7.99

From all over the dusty U.S./Mexican border come hair-raising stories of modern day encounters with winged monsters of immense size and terrifying appearance. Further field sightings of similar creatures are recorded from all around the globe. What lies behind these weird tales? Ken Gerhard is a native Texan, he lives in the homeland of the monster some call 'Big Bird'. Ken's scholarly work is the first of its kind. On the track of the monster, Ken uncovers cases of animal mutilations, attacks on humans and mounting evidence of a stunning zoological discovery ignored by mainstream science. Keep watching the skies!

STRENGTH THROUGH KOI
They saved Hitler's Koi and other stories

Jonathan Downes - ISBN 978-1-905723-04-1

£7.99

Strength through Koi is a book of short stories - some of them true, some of them less so - by noted cryptozoologist and raconteur Jonathan Downes. The stories are all about koi carp, and their interaction with bigfoot, UFOs, and Nazis. Even the late George Harrison makes an appearance. Very funny in parts, this book is highly recommended for anyone with even a passing interest in aquaculture, but should be taken definitely *cum grano salis*.

CFZ PRESS, MYRTLE COTTAGE, WOOLSERY, BIDEFORD, NORTH DEVON, EX39 5QR

Other books available from
CFZ PRESS

BIG CATS IN BRITAIN YEARBOOK 2007
Edited by Mark Fraser - ISBN 978-1-905723-09-6

£12.50

People from all walks of life encounter mysterious felids on a daily basis, in every nook and cranny of the UK. Most are jet-black, some are white, some are brown; big cats of every description and colour are seen by some unsuspecting person while on his or her daily business. 'Big Cats in Britain' are the largest and most active research group in the British Isles and Ireland. This book contains a run-down of every known big cat sighting in the UK during 2006, together with essays by various luminaries of the British big cat research community.

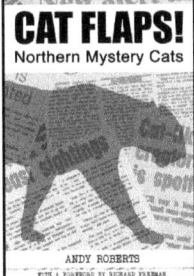

CAT FLAPS! Northern Mystery Cats
Andy Roberts - ISBN 978-1-905723-11-9

£6.99

Of all Britain's mystery beasts, the alien big cats are the most renowned. In recent years the notoriety of these uncatchable, out-of-place predators have eclipsed even the Loch Ness Monster. They slink from the shadows to terrorise a community, and then, as often as not, vanish like ghosts. But now film, photographs, livestock kills, and paw prints show that we can no longer deny the existence of these once-legendary beasts. Here then is a case-study, a true lost classic of Fortean research by one of the country's most respected researchers.

CENTRE FOR FORTEAN ZOOLOGY 2007 YEARBOOK
Edited by Jonathan Downes and Richard Freeman
ISBN 978-1-905723-14-0

£12.50

The Centre For Fortean Zoology Yearbook is a collection of papers and essays too long and detailed for publication in the CFZ Journal *Animals & Men*. With contributions from both well-known researchers, and relative newcomers to the field, the Yearbook provides a forum where new theories can be expounded, and work on little-known cryptids discussed.

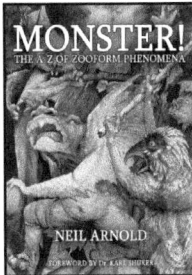

MONSTER! THE A-Z OF ZOOFORM PHENOMENA
Neil Arnold - ISBN 978-1-905723-10-2

£14.99

Zooform Phenomena are the most elusive, and least understood, mystery `animals`. Indeed, they are not animals at all, and are not even animate in the accepted terms of the word. Author and researcher Neil Arnold is to be commended for a groundbreaking piece of work, and has provided the world's first alphabetical listing of zooforms from around the world.

**CFZ PRESS, MYRTLE COTTAGE,
WOOLFARDISWORTHY BIDEFORD,
NORTH DEVON, EX39 5QR
w w w . c f z . o r g . u k**

Other books available from
CFZ PRESS

BIG CATS LOOSE IN BRITAIN
Marcus Matthews - ISBN 978-1-905723-12-6

£14.99

Big Cats: Loose in Britain, looks at the body of anecdotal evidence for such creatures: sightings, livestock kills, paw-prints and photographs, and seeks to determine underlying commonalities and threads of evidence. These two strands are repeatedly woven together into a highly readable, yet scientifically compelling, overview of the big cat phenomenon in Britain.

DARK DORSET
TALES OF MYSTERY, WONDER AND TERROR
Robert. J. Newland and Mark. J. North
ISBN 978-1-905723-15-6

£12.50

This extensively illustrated compendium has over 400 tales and references, making this book by far one of the best in its field. Dark Dorset has been thoroughly researched, and includes many new entries and up to date information never before published. The title of the book speaks for itself, and is indeed not for the faint hearted or those easily shocked.

MAN-MONKEY - IN SEARCH OF THE BRITISH BIGFOOT
Nick Redfern - ISBN 978-1-905723-16-4

£9.99

In her 1883 book, *Shropshire Folklore*, Charlotte S. Burne wrote: *'Just before he reached the canal bridge, a strange black creature with great white eyes sprang out of the plantation by the roadside and alighted on his horse's back'*. The creature duly became known as the `Man-Monkey`.

Between 1986 and early 2001, Nick Redfern delved deeply into the mystery of the strange creature of that dark stretch of canal. Now, published for the very first time, are Nick's original interview notes, his files and discoveries; as well as his theories pertaining to what lies at the heart of this diabolical legend.

EXTRAORDINARY ANIMALS REVISITED
Dr Karl Shuker - ISBN 978-1905723171

£14.99

This delightful book is the long-awaited, greatly-expanded new edition of one of Dr Karl Shuker's much-loved early volumes, *Extraordinary Animals Worldwide*. It is a fascinating celebration of what used to be called romantic natural history, examining a dazzling diversity of animal anomalies, creatures of cryptozoology, and all manner of other thought-provoking zoological revelations and continuing controversies down through the ages of wildlife discovery.

**CFZ PRESS, MYRTLE COTTAGE,
WOOLFARDISWORTHY BIDEFORD,
NORTH DEVON, EX39 5QR
w w w . c f z . o r g . u k**

Other books available from
CFZ PRESS

DARK DORSET CALENDAR CUSTOMS
Robert J Newland - ISBN 978-1-905723-18-8

£12.50

Much of the intrinsic charm of Dorset folklore is owed to the importance of folk customs. Today only a small amount of these curious and occasionally eccentric customs have survived, while those that still continue have, for many of us, lost their original significance. Why do we eat pancakes on Shrove Tuesday? Why do children dance around the maypole on May Day? Why do we carve pumpkin lanterns at Hallowe'en? All the answers are here! Robert has made an in-depth study of the Dorset country calendar identifying the major feast-days, holidays and celebrations when traditionally such folk customs are practiced.

CENTRE FOR FORTEAN ZOOLOGY 2004 YEARBOOK
Edited by Jonathan Downes and Richard Freeman
ISBN 978-1-905723-14-0

£12.50

The Centre For Fortean Zoology Yearbook is a collection of papers and essays too long and detailed for publication in the CFZ Journal *Animals & Men*. With contributions from both well-known researchers, and relative newcomers to the field, the Yearbook provides a forum where new theories can be expounded, and work on little-known cryptids discussed.

CENTRE FOR FORTEAN ZOOLOGY 2008 YEARBOOK
Edited by Jonathan Downes and Corinna Downes
ISBN 978 -1-905723-19-5

£12.50

The Centre For Fortean Zoology Yearbook is a collection of papers and essays too long and detailed for publication in the CFZ Journal *Animals & Men*. With contributions from both well-known researchers, and relative newcomers to the field, the Yearbook provides a forum where new theories can be expounded, and work on little-known cryptids discussed.

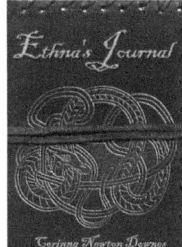

ETHNA'S JOURNAL
Corinna Newton Downes
ISBN 978 -1-905723-21-8

£9.99

Ethna's Journal tells the story of a few months in an alternate Dark Ages, seen through the eyes of Ethna, daughter of Lord Edric. She is an unsophisticated girl from the fortress town of Cragnuth, somewhere in the north of England, who reluctantly gets embroiled in a web of treachery, sorcery and bloody war...

**CFZ PRESS, MYRTLE COTTAGE,
WOOLFARDISWORTHY BIDEFORD,
NORTH DEVON, EX39 5QR
www.cfz.org.uk**

Other books available from
CFZ PRESS

ANIMALS & MEN - Issues 11 - 15 - The Call of the Wild
Jonathan Downes (Ed) - ISBN 978-1-905723-07-2

£12.50

Since 1994 we have been publishing the world's only dedicated cryptozoology magazine, *Animals & Men*. This volume contains fascimile reprints of issues 11 to 15 and includes articles covering out of place walruses, feathered dinosaurs, possible North American ground sloth survival, the theory of initial bipedalism, mystery whales, mitten crabs in Britain, Barbary lions, out of place animals in Germany, mystery pangolins, the barking beast of Bath, Yorkshire ABCs, Molly the singing oyster, singing mice, the dragons of Yorkshire, singing mice, the bigfoot murders, waspman, British beavers, the migo, Nessie, the weird warbling whatsit of the westcountry, the quagga project and much more...

IN THE WAKE OF BERNARD HEUVELMANS
Michael A Woodley - ISBN 978-1-905723-20-1

£9.99

Everyone is familiar with the nautical maps from the middle ages that were liberally festooned with images of exotic and monstrous animals, but the truth of the matter is that the *idea* of the sea monster is probably as old as humankind itself.

For two hundred years, scientists have been producing speculative classifications of sea serpents, attempting to place them within a zoological framework. This book looks at these successive classification models, and using a new formula produces a sea serpent classification for the 21st Century.

CENTRE FOR FORTEAN ZOOLOGY 1999 YEARBOOK
Edited by Jonathan Downes and Corinna Downes
ISBN 978 -1-905723-24-9

£12.50

The Centre For Fortean Zoology Yearbook is a collection of papers and essays too long and detailed for publication in the CFZ Journal *Animals & Men*. With contributions from both well-known researchers, and relative newcomers to the field, the Yearbook provides a forum where new theories can be expounded, and work on little-known cryptids discussed.

CENTRE FOR FORTEAN ZOOLOGY 1996 YEARBOOK
Edited by Jonathan Downes and Corinna Downes
ISBN 978 -1-905723-22-5

£12.50

The Centre For Fortean Zoology Yearbook is a collection of papers and essays too long and detailed for publication in the CFZ Journal *Animals & Men*. With contributions from both well-known researchers, and relative newcomers to the field, the Yearbook provides a forum where new theories can be expounded, and work on little-known cryptids discussed.

**CFZ PRESS, MYRTLE COTTAGE,
WOOLFARDISWORTHY BIDEFORD,
NORTH DEVON, EX39 5QR
w w w . c f z . o r g . u k**

Other books available from
CFZ PRESS

BIG CATS IN BRITAIN YEARBOOK 2008
Edited by Mark Fraser - ISBN 978-1-905723-23-2

£12.50

People from all walks of life encounter mysterious felids on a daily basis, in every nook and cranny of the UK. Most are jet-black, some are white, some are brown; big cats of every description and colour are seen by some unsuspecting person while on his or her daily business. 'Big Cats in Britain' are the largest and most active research group in the British Isles and Ireland. This book contains a run-down of every known big cat sighting in the UK during 2007, together with essays by various luminaries of the British big cat research community.

CFZ EXPEDITION REPORT 2007 - GUYANA
ISBN 978-1-905723-25-6

£12.50

Since 1992, the CFZ has carried out an unparalleled programme of research and investigation all over the world. In November 2007, a five-person team - Richard Freeman, Chris Clarke, Paul Rose, Lisa Dowley and Jon Hare went to Guyana, South America. They went in search of giant anacondas, the bigfoot-like didi, and the terrifying water tiger.

Here, for the first time, is their story...With an introduction by Jonathan Downes and forward by Dr. Karl Shuker.

CENTRE FOR FORTEAN ZOOLOGY 2003 YEARBOOK
Edited by Jonathan Downes and Corinna Downes
ISBN 978 -1-905723-19-5

£12.50

The Centre For Fortean Zoology Yearbook is a collection of papers and essays too long and detailed for publication in the CFZ Journal *Animals & Men*. With contributions from both well-known researchers, and relative newcomers to the field, the Yearbook provides a forum where new theories can be expounded, and work on little-known cryptids discussed.

CFZ PRESS, MYRTLE COTTAGE,
WOOLFARDISWORTHY BIDEFORD,
NORTH DEVON, EX39 5QR
www.cfz.org.uk

www.ingramcontent.com/pod-product-compliance
Ingram Content Group UK Ltd.
Pitfield, Milton Keynes, MK11 3LW, UK
UKHW021320180426
11947UKWH00015B/1333